Lightplane Maintenance

AIRCRAFT ENGINE OPERATING GUIDE

Lightplane Maintenance

AIRCRAFT ENGINE OPERATING GUIDE

Belvoir Publications/Kas Thomas

TAB Books
Division of McGraw-Hill
New York San Francisco Washington, D.C. Auckland Bogotá
Caracas Lisbon London Madrid Mexico City Milan
Montreal New Delhi San Juan Singapore
Sydney Tokyo Toronto

18442663

ALSO BY KAS THOMAS:

Personal Aircraft Maintenance (McGraw-Hill, 1981)
The Complete Book of Homebuilt Rotorcraft, with Jack Lambie (TAB)
How to Fly Helicopters, with Larry Collier (TAB)

This 1988 edition approved by the TAB aviation editorial staff.

© 1985 by **Belvoir Publications**
Published by TAB Books.
TAB Books is a division of McGraw-Hill

pbk   10 11 12 13 14 15 16 FGR/FGR 9 9 8

Library of Congress Cataloging-in-Publication Data

Thomas, Kas.
   Lightplane maintenance : aircraft engine operating guide / by
Belvoir Publications/Kas Thomas.
      p.   cm.
   Includes index.
   ISBN 0-8306-2431-7 (pbk.)
   1. Airplanes, Private—Motors. 2. Private flying. I. Belvoir
Publications. II. Title. III. Title: Light plane maintenance.
TL701.1.T54   1988
629.134′352—dc19   88-27561
                                                            CIP

Book Design by Carolyn A. Magruder.

Contents

Preface

Piston aircraft engines are a special breed. Most pilots are prone to regard today's Lycoming and Continental designs as quaint "period pieces," technological fossils-in-the-making. And it is undeniably true that piston aircraft engine manufacturing is a seasoned technology. (Lycoming and Continental have each been making reciprocating airplane engines for more than half a century—a remarkable fact, given that aviation itself is not even 90 years old.)

It is also true, however, that FAA-certificated piston engines are uncannily reliable by comparison with a variety of more "modern" types of machinery, such as motorcycle engines, lawnmowers, chainsaws, and ultralight powerplants, to say nothing of certain otherwise-trustworthy automobile engines which, when they've been tried in aircraft, have (how shall we say?) failed to instill confidence. It is probably no accident that most pilots would rather drive behind a Toyota engine, and fly behind a Lycoming, than vice versa.

But paradoxically, the very reliability of "recip" av-engines has led many pilots to become lax with regard to basic power management. Half the pilots you meet at the local airport can't tell you how to find an engine's leanest cylinder by reference to a multi-channel EGT gauge. (The other half can't tell you how to start a fuel-injected engine without flooding it.) A great many private pilots—even some flight instructors—don't know how a magneto works, or what the components of a "shower of sparks" ignition system are; and a lot of old-timers who should know better can't tell you why a runup is done at 1,800 rpm instead of idle or full throttle. It has only been in the last few years, really—with the recent surge in fuel prices and the exorbitant rise in overhaul costs—that the average pilot has begun to take more than casual interest in how his or her engine works, and how it should be flown.

To some extent, the appalling FWF ignorance shown by today's pilots is the result of Federal Aviation Regulations which have not kept pace with the increasing sophistication of small planes over the last 20 years. Under Part 61 of the FARs, as currently worded, it is perfectly legal (and quite common, of course) for a person to take the private-pilot check ride after having had *no exposure whatsoever* to: a manifold-pressure gauge; EGT instrumentation; constant-speed propeller operation; fuel injection; "alternate air"; auxiliary electric fuel pumps (high-wing Cessna trainers don't have such a thing); fuel-flow gauges; turbocharging; or cowl flaps. Some private pilots have not even seen a

CHT gauge by the time they've gotten their licenses. (Many training aircraft lack a cylinder-head-temp gauge.) And yet, over the last 20 years, the proportion of fuel-injected, cowl-flapped, constant-speed-propeller, and/or turbocharged aircraft in the civil fleet has grown enormously. Also, of course, the *cost* of those aircraft has grown enormously—which means that pilots literally can no longer *afford* to be ignorant on the subject of powerplant operation.

Unfortunately, while there are many books on how to fly in weather, how to fly "safely," how to fly at night, etc., there are virtually none on basic engine operation. The manufacturer's handbooks are admirably concise and to-the-point on this subject, but by the same token they rarely go into the kind of detail necessary to impart a genuine understanding of *why* an engine is leaned a certain way, or operated at a certain rpm, or started in a certain manner, etc. The manufacturer's manuals tell you *what* to do, but not *why* to do it.

This book is an attempt to tell both how *and* why aircraft engines are operated the way they are. It is aimed at all pilots, of all experience levels, who are interested in getting optimum performance and longevity from their engines. Since the text deals less with maintenance (per se) than with operational procedures, discussions of engineering matters are limited to just the amount of detail needed for an adequate understanding of the technique(s) in question. Readers who desire an expanded discussion of technical matters are encouraged to consult the excellent 554-page Fourth Edition of *Aircraft Powerplants* by Ralph D. Bent and James L. McKinley (1978, McGraw-Hill), or the FAA's equally compendious (if less up-to-date) *Airframe & Powerplant Mechanics' Powerplant Handbook* (AC 65-12A; Govt. Printing Office, 1976). In addition, the "Engine Clinic" department of *LIGHT PLANE MAINTENANCE* explores technical matters on a regular basis.

In writing a book of this kind, it is hard to decide on an appropriate level of "conversation" with regard to the (admittedly somewhat esoteric) subject matter. Pilots are a diverse group, after all. Some are technically trained, while others are not accustomed to talking in terms of "brake mean effective pressure" or BTUs. Some pilots, likewise, spend the majority of their flying hours in Cessna Skyhawks or other carbureted, normally aspirated airplanes with fixed-pitch propellers; others fly fuel-injected twins or high-performance singles. The temptation in a situation like this is for an author to find the lowest common denominator in his audience, and write for that group, keeping every discussion kindergarten-simple. At the risk of losing a few readers, I have decided *not* to reduce everything to monosyllables. Rather than start at "ground zero," I will assume that the reader already has some familiarity (perhaps through the automotive world) with piston engines, and that such terms as "camshaft" and "float carburetor" already have some meaning. (If not, there's a Glossary at the tail end of this book.) Students—and readers with very little technical background—may thus find some sections puzzling or fuzzy, for which I apologize. When complex matters are oversimplified rather than merely simplified, the truth often suffers (and the technically adept reader is snubbed). I wished to torture neither the truth nor the adept reader.

My first flying lesson was in 1962, and in a sense it could be said that this book was 20-plus years in the making. Over the years, I have had a chance to fly engines from just about all of the major "flat four" and "flat six" families, and I have had many excellent instructors. But most of what I know about aircraft engines has come from conversations over the years with just a few people—people such as Charles Melot of Mattituck (now Van Dusen); Roger Wulf, James L. Tubbs, and Gary Greenwood of Engine Components, Inc.; Joe Diblin of Avco Lycoming; Carl Goulet of Teledyne Continental; Charles E. Shader of Bendix; Al Hundere of Alcor; Paul Morton (retired Braniff captain and founder of the Cessna Owner Organizations); and the late Hugh MacInnes of Roto-Master. I am also deeply indebted to my father, Robert E. Thomas, whose advice on engine matters (as with so much else) has proven invaluable, and the many A&Ps—in California, North Carolina, Connecticut, and elsewhere—who have guided my maintenance forays in the past.

Special thanks are also due Carolyn A. Magruder (for her tireless willingness to follow through countless changes in typesetting and layout) and Rosemary Royer (for expert assistance with typesetting and proofreading). Barbara Hackett also provided critical typesetting services at a time when they were desperately needed. Dave Noland provided comic relief, thoughtful exchanges of ideas, and, frequently, deli runs. The writing of the book commenced in August 1984 and eventually consumed some 472K of disk space on an Apple Macintosh before being modem-dumped to a CompuGraphic 7500.

If this book, by heightening your awareness of what goes on ahead of the firewall, contributes to your enjoyment of flying, it will have served its purpose. If it helps your engine make TBO, so much the better.

Kas Thomas
Stamford, Connecticut
February 12, 1985

CHAPTER ONE
ENGINE DESIGN
BASICS

Engine Design Basics

Until World War Two, most of the popular aircraft engines in the U.S. were of the *radial* configuration—i.e., arranged with cylinders pointing away from the crankshaft like spokes around a hub (a configuration that owed its popularity partly to the fact that such engines are easy to air-cool, and partly to the fact that radials allowed compact, lightweight, easy-to-fabricate crankshaft and crankcase designs). In the course of World War Two, however, Lycoming and Continental—each of which had produced radial engines by the thousands—began mass-producing smaller aircraft engines in the now-familiar "opposed" layout, with cylinders parallel to each other and opposing each other, on opposite sides of a long (and somewhat flexible) crankshaft. For reasons having to do with vibrational characteristics, aerodynamics (cowl design), and aesthetics, opposed engines quickly became favored for applications under 300 horsepower, finding their way into a wide variety of postwar light planes, while radials were relegated to transport aircraft (only to disappear a few years later with the arrival of jets).

Today's Lycomings and Continentals actualiy differ very little, in overall layout, from the O-290s and E-185s of the Truman years. The "bottom end" of a modern reciprocating aircraft engine—which is to say, the crankshaft, crankcase, camshaft, and gearing (everything except the cylinder assemblies)—consists of a four- or six-throw (or even an eight-throw) steel-alloy crank, along with a steel-alloy cam, in a cast-aluminum crankcase with accessory gearing (oil-pump gears, magneto gears, vacuum pump pads, etc.) on the back. The "top end" components consist of steel cylinder barrels screwed or interference-fitted into cast aluminum cylinder heads, with valves riding in aluminum-bronze or cast-iron guides. The valves (which may be parallel or angled slightly, depending on the exact engine model) are pushrod-actuated. Ignition is provided, of course, by high-tension magnetos with mechanical breaker points.

Most pilots' understanding of how aircraft engines are designed ends here. Unfortunately, that's not good enough. Anyone who has any intention of getting maximum performance, reliability, and longevity from an aircraft engine (a high-performance engine, par-

Continental IO-520 (injected, opposed, 520 cubic inches' displacement) has 8.5-to-1 compression ratio, develops 285 horsepower, max-continuous. (Continental photo)

ticularly) owes it to himself or herself to know more than just the bare minimum about how the engine in question is put together. I would urge every pilot to make it a point, someday, to sit down with the appropriate engine manufacturer's overhaul or shop manual and study it awhile. In the meantime, you should at least familiarize yourself with the following anatomical details (which will be important in discussions throughout this book):

Crankcase. Probably one of the least understood (and most villainized) components of the modern air-cooled aircraft engine is the crankcase. All Lycoming and Continental cases are cast-aluminum, and as one sage was heard to comment, "anything made of cast aluminum will, sooner or later, crack." (*Why* cases crack is a discussion unto itself. Suffice it to say, if weight weren't a consideration in aircraft-engine design, aircraft crankcases would be made of cast iron and be twice as thick as they are now, and cracking would not be a problem. But that's not the way it is.) Aircraft crankcases may be sandcast—i.e., cast in sand molds—or cast in a permanent mold. Permanent-mold castings are typically used when it's necessary to obtain higher mechanical properties, better surfaces, and/or more accurate dimensions than are possible with sandcasting.

In theory, permanent-mold cases are stiffer, less porous, and generally more reproducible than sandcast cases, and thus afford the engine designer a chance to save weight, without compromising strength, through close-tolerance design. On a large, complex, potentially very flexible structure (such as a Continental GTSIO-520 crankcase), these considerations are of paramount importance—which is why Continental indeed uses the Permold process on most of its largest engines. Most other General Aviation engines, however, have sandcast cases, cast not by Lycoming and Continental themselves but by outside vendors. Lycoming and Continental take the "green" castings through complex final machining operations (now mostly computer-controlled), installing studs and Helicoils, etc., before sending the finished case to the assembly line. Among other finishing steps, the case halves are individually drilled to provide lubrication *galleries* for the flow of oil to moving parts (which we'll talk about more when we get to the discussion on "lubrication"). The crankcase also is ported to provide external ventilation via a long tube (called the *breather*). In addition, bosses are provided for special alloy-lined crankshaft bearings (or "main bearings"), which come in the form of separate, removable semicircular inserts. The inserts are replaced at each major overhaul.

Crankshaft. Aircraft-engine crankshafts are forged steel alloy (as opposed to many automotive cranks, which are cast-iron). If you ever see one by itself, you can't fail to be impressed with its beefiness. Still, an av-crank *will* break if abused. The sheer length of an opposed-engine crankshaft makes it unavoidably limber (particularly six- and eight-throw crankshafts, which, accordingly, often incorporate special pendulum-type dampers to dampen out harmful torsional—or twisting—vibrations). The main bearing journals, as well as the crankpins (i.e., the portions of the crankshaft that join to the connecting rods), are machined to ultra-close tolerances and nitrided (specially hardened) for long wear. In a four-cylinder Lycoming or Continental, the crankpins are oriented 180 degrees apart; in a six, the throws are 120 degrees apart. What's more, one bank (or row) of cylinders is always displaced just slightly ahead of the opposite bank, so that each piston has its own "fully dedicated" crankpin. (Again, in the automotive world, things are often done differently.) In a Lycoming engine, the cylinder numbering system is related in a straightforward way to the crankshaft layout: The number-one cylinder is defined as the one served by the forwardmost crankpin (i.e., the crank throw nearest the propeller). That can be either the front left or the front right cylinder, depending on the exact engine model. Continental, by

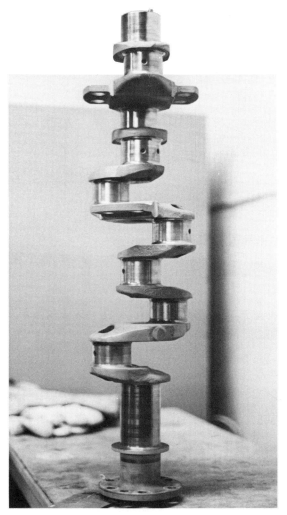

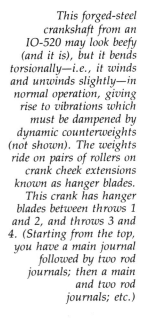

This forged-steel crankshaft from an IO-520 may look beefy (and it is), but it bends torsionally—i.e., it winds and unwinds slightly—in normal operation, giving rise to vibrations which must be dampened by dynamic counterweights (not shown). The weights ride on pairs of rollers on crank cheek extensions known as hanger blades. This crank has hanger blades between throws 1 and 2, and throws 3 and 4. (Starting from the top, you have a main journal followed by two rod journals; then a main and two rod journals; etc.)

contrast, defines the number-one cylinder as the cylinder served by the *rearmost* crankpin, which just happens to always be the right rear cylinder.

Crankshaft dampers. Many aircraft engine crankshafts, particularly on six and eight-cylinder engines, incorporate bifilar (dual-roller) pendulum-type counterweights for vibration damping. (Smaller engines, with short, stiff cranks, don't need dampers.) The counterweights consist of semicircular hunks of metal with two holes drilled out. They ride on individual crank cheek extensions known as *hanger blades* (also with two holes drilled out). A pair of roller pins, smaller in diameter than the holes they fit in, serve to connect the counterweights with the crankshaft hangers. The natural frequency of the pendulum weights is determined either by the mass of the counterweight, or (more often) by the size of the rollers (or in some instances by the size of the bushings that are used to contain the rollers). The need for crankshaft dampers becomes apparent when you consider that in normal operation, the limber crank twists and untwists (or winds up, and unwinds) very slightly in the direction of rotation, kind of like a rubber band (a very stiff one, admittedly). The torsional pulsation is a normal consequence of the intermittent combustion going on in the cylinders. As the power pulses are transmitted from piston to connecting rod to crankpin, the crankshaft itself responds by twisting and turning. In a six-cylinder, four-stroke-cycle engine, there are, of course, three power pulses per revolution of the crank. Accordingly, the principal modes of crank vibration in such an engine are multiples of three: third-order, sixth-order, etc. (The term "third order" refers to any vibration that peaks three times per revolution. When your car's front tires are out of balance, you experience two first-order vibrations—one from each side of the car.) If the engine designer puts counterweights on the crankshaft and "tunes" the counterweights just right, much of this type of vibration can be cancelled out, allowing smoother operation and longer engine life. (Conversely, if sludge builds up in the counterweight rollers, or the rollers are damaged in the course of a prop strike, or the wrong rollers are used during an overhaul, the counterweights will be *detuned* and the engine will run rough in certain rpm ranges.) The purpose of this discussion is not to make you an expert on engine vibrations (although you now know more about the subject than most A&P mechanics), but to familiarize you with the concept of dynamic counterweights, so that the idea of counterweight tuning and detuning doesn't remain a total mystery. (For more information on the subject, I recommend you refer to Lycoming Service Bulletin No. 245B.)

Connecting Rods. Aircraft engines employ automotive-type forged steel connecting rods with bronze piston-pin bushings at the piston end, and semicircular precision bearing inserts (the same type as make up the crankshaft main bearings) at the crankpin end. Lycoming and Continental do not make their own rod forgings, but obtain rod blanks from a forging vendor (such as Fiat), then subject them to shotpeening (to remove forging scale), machining, various quality inspections, and—finally—weighing, before assembling them to an engine. (Rods are weighed and separated into groups at the factory, to keep the weights within a half ounce in opposite bays. Later, the fully assembled crankshaft—complete with rods and dynamic counterweights—is dynamically balanced before engine final assembly takes place.)

Pistons. Pistons may be either cast aluminum or (in high-output engine models) forged aluminum, or in some cases forged aluminum with a steel insert inthe head portion to accept the top compression ring (e.g., Continental TSIO-520-AE and O-470-U, Spec. 18). Most Lycoming and Continental engines employ four-ring pistons. I.e., pistons are grooved to accept (from top down) two compression rings, an oil control ring, and—below the piston pin—a wiper ring. Many engines, however (including various Lycomings and older O-470 Continentals) use three-ring pistons, in which case the wiper ring is nonexistent. There are even a few engines with five-ring pistons, but these have mostly been updated to the four-ring configuration. Piston pins are full-floating steel tubes, usually with aluminum end plugs. Lycoming matches pistons by weight (plus or minus a quarter-ounce in opposite bays), but Continental stopped matching pistons years ago, after it was found that a weight difference of as much as half a pound in opposing pistons produced no abnormal performance in the course of a test-cell vibration survey. Once again, it should be noted that pistons—like most other engine components—are not actually manufactured by Lycoming or Continental, but come from outside vendors (such as Metal Leve, S.A., of Brazil). The Lycoming and Continental factories merely machine and inspect the vendor-supplied castings or forgings before putting their imprimatur on them.

Piston rings. Piston rings for Continental and Lycoming engines are made by Perfect Circle, Koppers Company, and/or other vendors. Ductile cast-iron is now the preferred ring material (for compression rings), although plain cast-iron and other steels are also used. Depending on the application, the outer edge (or face) of the ring may be chromed, plasma-coated with molybdenum, or treated in other ways. Also, depending on piston-groove design, the ring cross-section may either be rectangular (e.g., Continental O-470-R), half-wedge (Lycoming TIO-540-J), or full-wedge (Lycoming O-235-C). Oil control rings are generally slotted, machined around the circumference to provide oil drain-back holes, and supported on the underside with a spring to provide constant tension. Wiper rings have an obvious bevel at the outer edge, but otherwise look and perform much the same as a compression ring.

Cylinders. In Lycoming as well as Continental engines, an externally finned, aluminum-alloy head casting is heated to about 600 degrees Fahrenheit (316 Celsius)—and valve inserts cold-pressed into

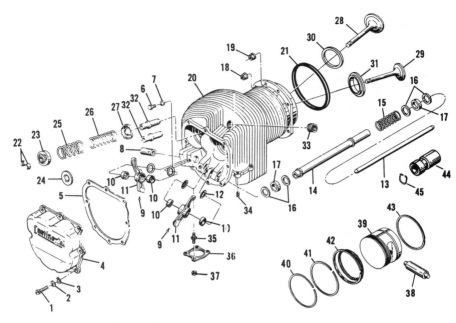

Cylinder and Piston Assembly

1. Screw, Fillister Head
2. Washer, Lock
3. Washer, Plain
4. Cover, Valve Rocker
5. Gasket, Valve Rocker Cover
6. Screw, Hex Head
7. Washer, Plain
8. Shaft, Valve Rocker
9. Screw, Drive
10. Bushing, Valve Rocker
11. Rocker
12. Washer, Thrust
13. Push Rod
14. Housing, Push Rod
15. Spring, Push Rod Housing
16. Washer, Push Rod Housing
17. Packing, Push Rod Housing
18. Nut, Flanged
19. Nut, Flanged
20. Cylinder Assembly
21. O-Ring, Cylinder Base
22. Key, Retainer
23. Roto Coil Assembly, Exhaust Valve
24. Retainer, Intake Valve
25. Spring, Outer
26. Spring, Inner
27. Retainer, Lower
28. Valve, Intake
29. Valve, Exhaust
30. Insert, Intake Valve
31. Insert, Exhaust Valve
32. Guide, Valve
33. Insert Spark Plug
34. Insert, Intake Flange
35. Stud
36. Gasket Assembly, Exhaust Flange
37. Nut, Hex Head
38. Pin and Plug Assembly
39. Piston
40. Ring, Compression
41. Ring, Compression
42. Ring Assembly, Oil Control
43. Ringer, Scraper
44. Lifter Assembly, Hydraulic Valve
45. Ring, Retaining

Air-cooled cylinders consist of a cast-aluminum (AMS 4225) head shrunk onto a steel barrel. On the left, a small-finned, parallel-valve jug from a Continental TSIO-520-AE; at right, a standard IO-520 cylinder. (Continental photo)

place—before the head is screwed, butted, and shrunk onto an externally finned steel alloy barrel to create what you and I call a cylinder assembly (or jug). Actually, the complete cylinder assembly also incorporates valves, valve springs (and rotators), rocker arms, rocker shaft(s), rocker cover (aluminum casting), pushrod housings, and pushrods, plus various studs and helical-coil thread inserts. Because the head end of the cylinder barrel is hotter than the crankcase end, it tends to expand or mushroom out in normal operation; for this reason, aircraft cylinder barrels are *choked*, or ground to a tapered bore (smaller at the top). The amount of taper is not much—perhaps seven or eight thousandths of an inch, on a diameter (or bore) of five inches—but it is important, nonetheless, in the context of oil consumption and ring wear. After grinding for choke and dimension, cylinders are honed to a specific crosshatch pattern and scratch depth before being put in service. (In addition, the barrel may or may not be nitride-hardened. If it's a remanufactured barrel, it may be chrome-plated on the inside to restore dimensions to new or service limits. Look for a band of red or red-orange paint at the base of the cylinder.)

Camshaft. Lycoming and Continental cams are forged steel alloy billets which are machined on journal surfaces, cam lobes, and the gear mounting flange at the rear end. (On Lycomings, one end may also be machined to form a fuel-pump-actuating cam.) In Permold-case Continentals, the hollow camshaft serves as the engine's main oil gallery. Generally, each exhaust lobe operates one exhaust valve, whereas intake lobes (which have a larger ramp area) operate two opposing intake valves. Thus, a four-cylinder engine will have a six-lobe camshaft (four exhaust, two intake lobes) and a six-cylinder engine will have a nine-lobe cam. One important difference between Lycoming and Continental engines is that Lycoming always locates the camshaft *above* the crankshaft, along the spine of the engine, whereas in Continental engines, the cam sits *below* the crank. Some people feel the "high and dry" location of the cam in Lycoming engines contributes to a greater incidence of cam distress (caused by oil runoff?) in Avco products, but such a contention is difficult to prove conclusively.

Tappets. Except for certain Lycoming O-235 and O-290 series engines, all current designs use hydraulic tappets (or lifters) for valve actuation. (The O-235 and O-290 use solid tappets.) Continental's tappets are of the familiar (automotive) barrel type, and may be removed and replaced without complete disassembly of the engine. (The tappets ride in special crankcase bosses, easily accessible from outside the engine.) Lycoming tappets, on the other hand, are mostly of the mushroom type (large end against the cam lobe) and cannot be completely removed from the engine without splitting the crankcase. An exception is made in the case of O-320-H and O-360-E Lycoming engines, which use barrel-type (i.e., easily removable) lifters. Regardless of configuration, all hydraulic lifters are designed to do the same thing: namely, provide for zero lash in the valve train. You can think of hydraulic lifters as small, spring-loaded oleo struts that cushion and take up slack in the mechanical circuit from cam lobe to pushrod to rocker to valve stem. (In a pushrod-type engine *without* hydraulic tappets, the differential expansion rates of the various engine components as the engine changes operating temperatures acts to create gaps between moving parts. Valve-train components then slap or hammer each other mercilessly during the reciprocating action of the valve. When a hydraulic tappet gets contaminated with dirt or deposits and sticks in the collapsed position, it acts like a solid tappet, and the result is usually heard as valve hammering.) Hydraulic tappets for aircraft use are made by the same vendors (e.g., Eaton Corporation) that make tappets for the automobile industry.

An intake valve and an exhaust valve from a Lycoming IO-360 engine. The intake valve (left) has a large face to allow easy inflow of air and fuel; the exhaust valve needn't be as large, since high-pressure combustion gases tend to flow out quickly. The exhaust valve shown here is a sodium-filled model listing for well over $200.

They come as complete assemblies, with a custom fit that is quite critical; it's important never to mix tappet plungers and bodies, as the bleed-down rate may be altered to the extent that valve hammering occurs.

Valves. Intake valves, because they are fuel-cooled, operate at a fairly low temperature and are thus usually made of low-temperature austenitic steel alloy (e.g., 'KB' steel), forged in one piece. Exhaust valves are a little different story: The flow of super-hot exhaust gases past the (open) exhaust valve has a blowtorch effect, which allows exhaust valves to reach temperatures of 1,600 degrees F or more. Special materials are clearly called for. Here once again, Lycoming and Continental part ways somewhat. Where Continental engines use solid-stemmed exhaust valves exclusively, Lycoming specifies sodium-cooled exhaust valves for all its current-production engines. Lycoming valves (actually made by Eaton Corporation) are hollow-stemmed and partially filled with elemental sodium. In normal operation, the sodium melts and sloshes back and forth, carrying heat from the head (hot part) of the valve to the stem (which has a large contact area with the guide and cylinder head). Sodium-filled valves dissipate heat better than their solid-stemmed counterparts, and normally operate about 200 degrees (F) cooler as a result. Nevertheless, special high-temperature alloys, such as Nimonic 80A or Inconel, must often be used in their construction; and for proper cooling to occur, good contact must be maintained between the exhaust valve stem and the valve guide throughout the service life of the engine. (The stem and guide, for all practical purposes, operate dry. If clearances open to the point where oil enters the guide, problems begin.) Continental has chosen to avoid sodium-cooled valves, for reasons having to do with cost (some Lycoming valves list at over $350 each, whereas Continental valves seldom cost half that much), simplicity (guide clearances are not as critical), and safety (hollow-stemmed valves are not as damage-resistant as solid-stemmed valves). Continental exhaust valves—manufactured by a subsidiary of TRW—are now almost exclusively made of Nimonic 80A high-temp nickel alloy, hard-chromed on the stem, nitrided at the tip, and Stellite-faced at the head (for good corrosion resistance).

Valve guides. Not very many years ago, all aircraft engine valve guides were babbitt-bronze or aluminum-bronze (many still are). With the recent trend towards chrome-plating of valve stems, a harder wear surface was needed on the guide I.D. (inside diameter), and cast-iron or Ni-Resist guides came into fashion. Around 1982, Continental began phasing in production of Nitralloy guides (nitride-hardened steel alloy) on its biggest engines, and now the Nitralloy guides are being recommended for all high-output Continentals down to the O-470s (provided the best, latest-P/N hard-chromed Nimonic exhaust valves are used). Old-style bronze guides can be used with chrome-stemmed, high-temp exhaust valves (and in fact, that's what you'll get in a cheap field overhaul), but it's not recommended, since rapid guide wear often occurs. To some extent, the type of exhaust valve and guide your engine has determines the aggressiveness with which you can operate the engine and still expect to reach TBO, so it's worth knowing what kinds of components you have. (Crosscheck your engine log or overhaul work order against the latest revision of Lycoming Service Instruction No. 1037 or Continental Service Bulletin No. M82-6, as appropriate.)

Turbochargers. All of General Aviation's highest-output engines are equipped with turbochargers (made either by Rajay or AiResearch). A turbocharger, of course, is nothing more than two air pumps on a common shaft. One pump is made of high-temperature nickel alloy and sits inside a snail-shaped housing (made of ductile Ni-Resist or stainless steel), where it is turned at fantastic speeds—often over 100,000 rpm—by engine exhaust gases. This exhaust pump (or turbine) in turn spins the common shaft (which rides in a journal lubricated by engine oil pressure) and drives an impeller which compresses outside air before passing it to the cylinders. The remarkable thing about modern turbochargers is how compact they are: the smallest Rajays weigh in at about 13 pounds, while the heftiest AiResearch blower tips the scales at a mere 33 pounds or so. The other remarkable thing is how hot they get. In normal operation, with the wastegate even partway closed, a turbocharger glows cherry-red. Despite the use of special alloys, the exhaust portion of the turbocharger is limited to 1,650 degrees Fahrenheit, so the turbo itself is often the limiting factor in how hard you can ''push'' an engine (or hard severely you can lean it).

Starter adapter. Here is still another department in which Lycoming and Continental design philosophies differ sharply. Lycoming engines, without apparent exception, transmit starter-motor torque to the crankshaft for starting via a Bendix drive unit, which couples to a large, flangelike ring gear at the front of the engine. The Bendix unit consists of a centrifugally activated, spring-loaded pinion gear which (through a clever spiral-splined-shaft arrangement) simply flies out to meet the crankshaft ring gear a split-second after the starter motor starts whirring. After the engine picks up speed—and begins to drive the starter—the Bendix pinion pops back out of place, and the pilot is on his or her merry way. (Many people have forgotten, or never knew in the first place, that this type of starter adapter—or one very similar—was what originally got the financial wheel of fortune started for Vincent T. Bendix back in the 1920s.) The Bendix drive is an impressively simple and lightweight affair, its chief drawback being that it sits more or less out in the open—exposed to dirt, rain, ice, etc.—at the front of the engine compartment, where it's likely to get crudded

Bendix starter adapter used on Lycoming engines is lightweight, simple, and reliable except when dirty or iced-over. Lower left portion couples to starter; upper right portion flies out to meet the prop ring gear automatically when starter revs, thanks to internal spiral-grooved shaft (not visible here).

up in a hurry. And Bendixes definitely do not work well when they've been crudded up. Dirt (or frozen grease, etc.) on the spiral-splined shaft can cause the sliding pinion not to slide, in which case the starter motor never couples to the engine (very bad for the starter motor's bearings, since the motor inevitably overrevs), or—as the case may be—it never *un*couples. Suffice it to say, Bendix drives need frequent (once every 50 hours, ideally) cleaning, lubrication, and inspection, for optimum (make that acceptable) performance.

The alternative to the Bendix drive is the Teledyne Continental starter adapter (invariably located at the rear of the engine and fully sealed—hence not easily contaminated with crud). Actually, there are two types of Continental starter adapter: that used on the O-200 (Cessna 150), and the clutch-drum type used, in one variation or another, on all Continentals other than the O-200. (In truth, there's also the manual-pinion-engagement "pull starter" used on very early Cessna 150, 172, and 175 models, but production of that model starter clutch ended around 1965, and most owners have long since had their planes converted over to more modern starter types for which parts are widely available.) The O-200 starter clutch deserves brief mention, both because it's a fine design and because it fails often. The O-200 starter adapter consists of what appears to be a short shaft with unequal-sized gears on either end. In reality, there is a shaft attached to one gear, and a concentric housing attached to the other, with a sprague clutch inside the housing, so that with the large gear held stationary the small one can be turned easily in one direction, but "grabs" tight in the other direction. Unfortunately, the tiny sprague blocks inside the housing can easily wear and/or fret the housing race or drive shaft to the point where all the "grab" is gone, and the clutch freewheels in both directions—in which case your starter won't turn the engine. The sprague housing receives oil lubrication from the engine oil system, and it would appear (from the many user reports filed over the years) that extra-careful attention to oil cleanliness, and choice of correct viscosity, are important for long starter-clutch life in this type of engine.

The standard Continental starter clutch, used on everything from the IO-360 up, has a Prestolite starter motor turning a worm gear which drives a ring gear on a rotating clutch drum. The drum abuts a shaftgear (which transmits torque to the engine) which has its own drum. Both drums are externally grooved and share a common spring (a heavy helical spring) which, when the starter is energized, tends to want to wind up on both drums and lock the two together. Torque is thus transmitted to the crankshaft until the engine fires. After the engine fires—and starts turning the shaftgear drum—the drums tend to unwind the spring, and the spring relaxes, allowing the two drums to demate. It sounds more complicated than it is (trust me). As it turns out, this type of starter adapter has proven remarkably reliable over the years, and Continental has stuck with it for all high-performance TCM engines.

Miscellaneous. Oil pans, accessory case covers, and rocker covers are aluminum castings. Oil pump gears may be steel alloy, sintered iron, or aluminum. (Lycoming had poor luck with sintered iron impellers; most have now been replaced under Airworthiness Directive 81-18-04.) Engine mounts are made of welded steel tubing and rarely crack or give problems, except when some misguided soul wraps asbestos cloth or tape around sections of tubing, trapping moisture underneath and causing rust-through. Corrosion is a problem in steel engine mounts, and it's important to keep the tubing painted and shielded from heat in hot spots. Engines may be held in their mounts in one of two ways depending on the exact model. Many smaller engines are cantilevered, with mounting lugs at the rear of the crankcase only; others are cradle-mounted (with lugs fore and aft, underneath the cylinders). Large rubber shock biscuits, specially designed, are placed at the attach points to absorb vibration. (The rubber shock mounts are mostly made by Lord Corporation, and are fairly expensive to replace—$300 or $400 a set, usually). When the mounts have been designed in such a fashion that imaginary lines, extended through the center of each shock biscuit, intersect at the center of gravity of the engine, the setup is known as a Dynafocal-type mounting.

Exhaust pipes and mufflers for General Aviation aircraft are usually (though not always) designed and supplied by the airframe manufacturer, rather than the engine manufacturer; technically, they are part of the airframe, not the engine (as far as the FAA is concerned, at least). We'll merely note here that most exhaust components, on most aircraft, are made of carbon steel. Some systems are fabricated from 321-type stainless; and a precious few employ Inconel parts (which are expensive, but extremely durable). Some after-market suppliers, such as Wall Colmonoy Corporation of Oklahoma City, offer replacement exhaust parts made of more durable materials than are found in many original components. (Colmonoy is famous, for example, for its Nicrocoat components, which receive a plasma spray-coating of high-nickel, corrosion-resistant superalloy.) The important point to remember is that heat and moisture are just as corrosive to aircraft exhaust pipes as they are to automobile mufflers,

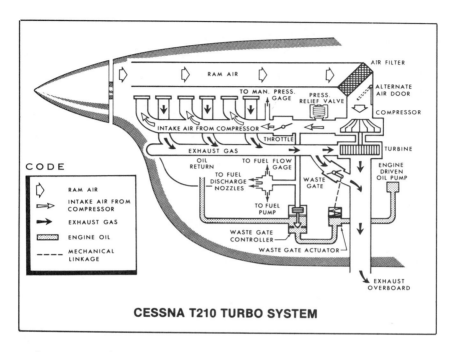

CESSNA T210 TURBO SYSTEM

only more so, because where cars operate at 20 or 30 percent power most of the time, aircraft engines generally operate throttle-to-the-wall. (Also remember that single-engine planes get cabin heat from air ducted around the outside of the muffler or exhaust pipe, so that the slightest exhaust pinhole can let noxious fumes into the cockpit.) The type of exhaust system you have will have (or should have) some bearing on how aggressively you lean your engine. The best exhaust components are Inconel (good for temperatures to 1,700 degrees); next-best, plasma-coated (Nicrocoat) steel; then stainless; then ordinary carbon steel. Of course, if you have a normally aspirated, low-output (under 200 horsepower) engine, your exhaust may never get hotter than 1,400 degrees in cruise—even if you lean to peak EGT—in which case a mild-steel exhaust system is just fine. If on the other hand you have a high-output engine with a plain-steel exhaust system, you'd be smart to keep a close eye on EGT in climb and cruise: Even 321 stainless encounters serious creep problems at 1,600 degrees. (Creep refers to a metal's tendency to become plastic and flow at high temperature.)

Summary. In stark contrast to recent developments in the automotive world, piston engines for aircraft today look much the same as they did forty years ago, with externally bolted-on cylinders, pushrod-actuated valves, and a relatively heavy forged-steel-alloy crankshaft enclosed in a cast-aluminum crankcase. What few changes have occurred over the years have been mainly in the choice of materials for valves, guides, pistons, rings, bearings, and exhaust components; also, compression ratios and takeoff rpm have risen somewhat. And turbochargers—which made their first appearance on lightplane engines in 1962—are now commonplace. But for the most part, we find ourselves flying radial engines that have been taken apart and reassembled with the components laid flat. (Fully disassembled, with all parts laid out on a work bench, it is hard to tell a radial from a modern-day opposed engine—except for the crankshaft and crankcase.)

Given the apparent lack of major design changes in aircraft engines over the past four decades, one might be tempted to assume that relatively few changes have occurred in the way engines are *flown*. Such is certainly not the case, however. As we'll see in the next chapter (and throughout this book), subtle changes in instrumentation, fuel delivery, cooling, and lubrication—and *major* changes in rpm, rated power, and airplane flight envelopes (not to mention engine prices)—dictate new ways of thinking about powerplant management. The need for pilot education in this area has, in fact, never been greater.

CHAPTER TWO
COCKPIT CONTROLS

Cockpit Controls

Since this book is about intelligent use of an airplane's powerplant controls—nothing more, nothing less—it is basic to our task to discuss the function of each of an airplane's go-knobs, one by one, starting more or less from scratch. Granted, if you've been flying high-performance piston aircraft steadily for the last twenty years, very little of what we have to say here will be new to you and you may rightfully skip on to the next chapter. If not, please be seated; the following information is crucial.

Sad to say, one of the purposes of this chapter is to undo some of the damage done by improper tutoring in the use of power controls by certified flight instructors (who in some cases have flown nothing but Tomahawks and 172s their whole lives). For better or worse, present Federal Aviation Regulations allow a student pilot to qualify for the private license after 40 hours in a Cessna 152, Tomahawk, or other carbureted, normally aspirated, fixed-pitch-prop, generally-low-in-performance nose-dragger. There is no requirement that private pilots learn anything at all about fuel injection, constant-speed props, manifold pressure, EGT, cowl flaps, turbocharging, or even boost pumps. (High-wing trainers generally don't have one.) Most of these controls are learned on a by-guess-or-by-gosh, trial and (pilot) error, "modified Indiana Jones" basis. (Let's see, how hard could this be?)

Precisely because training airplanes *are* so simple, oversimplified explanations of how power controls work are normal and acceptable in the early stages of training. (The beginning pilot certainly has better things to worry about than stoichiometric mixture control.) For the first 10 or 20 hours, it's enough to know that the throttle is like the foot pedal in a car, the key is the ignition (again like in a car), and the mixture is kind of like a glorified choke control which also happens to stop the engine at the end of every flight. Unfortunately, many pilots go on to add hundreds of hours to their logbooks before learning what these controls (and the others mentioned above) *really do*. Time to clear the cobwebs.

Master Switch

The master switch is not a powerplant control. (This is a trick section.) We include it in our discussion, however, for the simple reason that many pilots seem to think it *is* an engine-related switch—a belief that stems, apparently, from the fact that you can't *start* an aircraft engine (with an electric starter) without first turning on the master. Just so there'll be no confusion on the matter, let's review what happens:

When you engage the master switch, or *electrical master* (to distinguish it from the *avionics master* that many high-performance planes also have), you are supplying battery power to the plane's electrical system. In most small planes, the battery is directly wired to very few items that don't go through the master switch; perhaps some courtesy lights in the cabin area, an electric clock, a cigaret lighter, and/or an optional alarm system of some kind. To get anything else electrical to come on, you have to activate the master switch, the immediate consequence of which is that a *contactor* (or *relay*) adjacent the battery energizes, allowing full battery output to be made available to the plane's electrical bus or buses, plus the starter-motor circuit (which is isolated from the electrical bus due to the large amounts of current that must pass through the circuit). If you're not familiar with the term "bus" in this context, you can think of the bus (or "bus bar") as a sort of giant electrical outlet into which all of the plane's accessories plug.

The starter circuit is enabled whenever the master switch is turned on. But to actually energize the starter motor and crank the engine, you must press a "start" button, or turn and hold (or even push in on) the ignition key, thereby causing a relay near the starter motor (often called the starter *contactor*) to close, allowing the battery's full output to flow to the starter itself. (If the master switch is *off* as you hit the starter button or turn the key, the starter contactor will not make contact, and you will not see any starter action.) All of this is shown diagrammatically in the electrical system schematic that accompanies your plane's owner's manual. Be sure to study this schematic periodically, particularly when checking out in an unfamiliar airplane. There are subtle difference in bus bar connections, ground service receptacle wiring, night-light wiring, etc., between aircraft models.

Most newer planes have split master switches. In this setup, one half of the switch acts to send battery power to the bus bar and starter circuit, while the other half (marked 'ALT') sends battery current to the alternator field windings. (The electromagnetic field produced in the field windings are what enable an alternator rotor, upon turning, to produce electricity. Without a small current flowing through these windings, the alternator would be useless.) The idea here is that in the event of an inflight alternator failure, you can punch the 'ALT' half of the master off and prevent battery power from being needlessly wasted on energizing the dead alternator's field windings. (Also, if you're on the ground listening to the radio with the engine stopped, you can keep the 'ALT' switch off and conserve battery power—and prevent the alternator from heating up.) During engine startup, it's usually a wise move to keep the 'ALT' switch in the off position until the engine has picked up speed, both to save cranking power and to prevent voltage surges from possibly damaging the alternator. A word of caution, however: Should you choose to turn the 'ALT' portion of the master switch off temporarily in normal operation, remember not to run the battery so low that when you flick the 'ALT' switch back on, there's insufficient current available to energize the alternator field. Without several amperes to excite the field windings, your alternator will be produce zero output, regardless of rpm.

Ignition Switch

Ignition switches can either be very simple or quite complex, depending on the type of aircraft and magneto variety. The simplest arrangement, oddly enough, is often found in multiengine aircraft, which may use independent toggle switches for each magneto (along with separate starter buttons located elsewhere on or under the panel). Single-engine aircraft, by contrast, frequently employ a key-rotated ignition switch with five positions: off, left, right, both, and start. (The start position is invariably spring-loaded to return the key to "both" when finger pressure is released.) In some airplanes, a four-position key ignition (off, left, right, both) is utilized in conjunction with a remote-mounted starter button.

It's not important for you to understand the workings of your ignition switch(es) in greater detail at this point. (We'll have more to say about ignition in the following two chapters.) Nonetheless, you should at least be cognizant of the following "key facts":

1. Unless you have a "shower of sparks" ignition (see next chapter), and unless your switch is of the combination key-start kind, no portion of the ignition system in your aircraft is in any way connected to the aircraft electrical system, battery, buses, etc. As long as the ignition is on the left, right, or both position, and the engine is turning over, the magnetos will deliver electrical energy to the spark plugs regardless of what the airplane's alternator, generator, or battery are doing, or not doing. Your magnetos are "self-exciting"—they use permanent magnets (rather than battery-fed field windings) to generate juice. (Hence the name "magnetos," of course.)

2. Each magneto is wired to the ignition switch in such a way that

when the switch is off, the magneto is grounded through the switch. (Conversely, when the switch for that magneto is "on," the mag's primary lead—or P-lead—is ungrounded, and the magneto is said to be "hot.") If you think about it, you can see that turning *on* a magneto is the functional equivalent of snipping the magneto's primary lead with wire cutters, at the switch itself. Conversely, any time a P-lead breaks or becomes disconnected, the magneto in question becomes "hot" just as surely as if the ignition switch had been turned on (even if, in fact, it hasn't been). I bring this up because P-lead breakage is not uncommon, and—since aircraft engines are stopped by cutting the fuel, not the ignition—such a condition can escape detection for quite some time. That's why, any time you are handling a plane's propeller, you should treat the prop as if the engine is hot, even if the keys are in your pocket.

3. Ignition switches themselves have been known to fail, short-circuit, or be defective fresh out of the box, or be wired incorrectly to the magnetos. Always bear this in mind when it comes to troubleshooting an ignition-related problem.

Boost Pump

Not all planes have auxiliary electric fuel pump, or boost pump. In planes that do have one, the extra pump is usually (but not always) mounted in series with the engine-driven pump, on the upstream side (sometimes ahead of the firewall; other times—as in the author's Cessna 310—in the main fuel tanks themselves). Some airplanes, such as the Cessna 177B Cardinal, have a boost pump mounted *in parallel* with the engine-driven pump. Regardless of plumbing, the purpose of the auxiliary fuel pump—which is controlled by a switch on the panel—is to restore adequate fuel flow to the engine at those times when the engine-driven (mechanical) fuel pump, for whatever reason, is incapable of providing adequate fuel flow to sustain controlled combustion. In hot weather, or above 12,000 feet, the auxiliary fuel pump is primarily used to suppress vapor formation.

I might add that when the engine in question is fuel-injected rather than carbureted, the boost pump has another very important function: it enables the pilot to *prime* the engine prior to startup.

Boost pumps vary in design from one airframe installation to the next. (Note that it is the airframe manufacturer who, more often than not, specifies the boost pump design.) Some pumps are controlled by a two-position (actually a *three*-position) switch that allows for a choice of low or high output, in addition to "off." In most such cases, the pump's output when the "high" position is selected is so great as

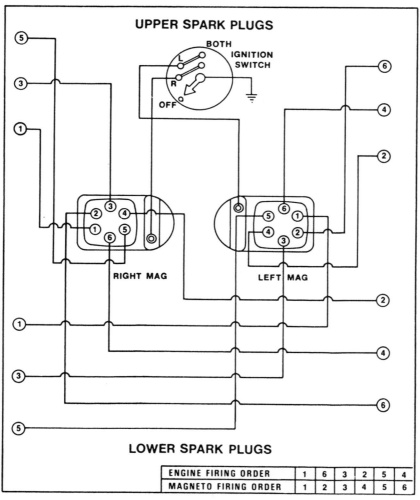

UPPER SPARK PLUGS

BOTH
IGNITION
SWITCH

OFF

RIGHT MAG

LEFT MAG

LOWER SPARK PLUGS

ENGINE FIRING ORDER	1	6	3	2	5	4
MAGNETO FIRING ORDER	1	2	3	4	5	6

A typical magneto installation has the left mag firing the top plugs in the odd-numbered cylinder and the bottom plugs in the even jugs; and vice versa for the right mag.

to flood the engine, or cause it to lose power from overrichness. In planes that have a simple on/off boost pump, by contrast, standard operating procedure is to leave the boost pump on for takeoff and landing, and whenever vapor formation is encountered (or merely anticipated). Obviously, you'll want to read your owner's manual carefully to determine how to use your boost pump properly; it's hard to generalize here, due to the many installation differences that exist from plane to plane.

Do remember this: The auxiliary fuel pump is an electrical accessory and therefore subject to loss of function during episodes of low or no voltage. Also, because it *is* an electrical accessory, you'll find a circuit breaker or fuse labelled "aux pump" on your plane's circuit breaker or fuse panel. Check this panel if you cannot ascertain from fuel-pressure indications whether or not the boost pump is functioning.

Fuel Selector

Although not an engine control in the usual sense, the fuel selector valve does have an attention-getting effect on engine power output if abused, so we'll discuss it briefly here. Most fuel selector handles are located on the floor or sidewall near the pilot's feet; but the selector valve itself need not be located near the handle. (In the Cessna 310, for example, the handles are on the floor, but the valves are out in the wings. Even in a 172, the handle is displaced a good 18 inches from the valve itself.) The main things to keep in mind about the fuel selector are:

1. Because the handle is often physically far removed from the valve itself, it's never wise to trust handle or pointer indications. (The mere fact that the handle is pointing to "left tank" doesn't automatically mean you're on the left tank.) Instead, you should *feel for a detent* every time you move the handle to a new position. If you don't feel a detent, don't trust the selector. Also, make it a habit to look at the selector handle *after* everyone has boarded the airplane. Passengers (and pilots) have a way of accidentally stepping on or kicking fuel selector handles during the final boarding process. (To get to the pilot's seat in a 310, you practically have to walk across both fuel selectors.) Every year, half a dozen pilots make off-airport landings with the fuel selector positioned between tanks. Don't be one of them.

2. Never turn a selector to "off." The exception to this rule is if you're experiencing an engine-compartment fire, or you're about to crash and you don't have anything better to do on "short final." The rest of the time, leave the fuel on. There is absolutely no valid reason to turn the fuel off between flights, even during extended periods of inactivity. It's too easy to forget to turn it back on later.

3. Read your Pilot's Operating Handbook, and *know your fuel system* before flipping the fuel selector to "aux tanks." (In the 310, for example, excess unmetered fuel is returned to the main tanks when operating on "aux" tanks, so it's necessary to fly an hour or more on the mains before aux fuel is selected—otherwise fuel returning to the already-full mains will be pumped overboard out the vent. Also,

there is no boost-pump action when flying on auxiliary fuel in a 310, since the boost pumps are inside the tip tanks.) I have a personal habit of subjecting fuel selectors to the least amount of use possible. If I'm in a plane that has a "both" setting, that's where I leave the fuel selector. I don't believe in flip-flopping back and forth constantly to achieve an even fuel burn in alternate wing tanks. Why wear out the handle? (Don't laugh. I know pilots who have done just that, including one man whose fuel handle sheared off in his hands—leaving the valve stuck between tanks—on a flight over mountainous terrain. He landed his Bonanza gear-up in a tight valley as a result.) Also, I'm a firm believer in using auxiliary tanks in straight-and-level flight only (having had the left engine of an Apache quit on me at low altitude during a letdown on aux fuel).

4. When engine power loss occurs, *hit the boost pump and switch tanks,* regardless of any other advice given above.

Mixture Control

The mixture control is certainly one of the three most important powerplant controls on the panel (the other two most important ones being, of course, the throttle and prop-pitch controls). In most planes, it consists of a red-painted lever or push-pull knob similar to (but smaller than) the throttle and displaced just to the right of and/or below the throttle and prop controls. The cockpit lever connects to the carburetor or fuel injector body via a long, flexible metal cable. When the pilot moves this cable, a tiny arm on the side of the carburetor or fuel injector (near where the fuel enters the unit) also moves, varying the amount of fuel that is allowed to pass through the carburetor's or fuel injector's main metering jets or orifices. Thus, you can think of the mixture control as a kind of master cutoff valve (infinitely adjustable) governing fuel flow to the engine.

Because the mixture control adjusts the final fuel flow in an absolute fashion, independent of air flow, it gives the pilot precise control over the fuel-air ratio, or "mixture," available to the cylinders. With the mixture knob pulled all the way aft, into a position known as *idle cutoff,* no fuel at all is allowed to escape the carburetor or fuel injector. In this position, the fuel-air mixture is said to be "full lean." The full-forward position, by contrast, is often called full *rich*—because in that position, the fuel-air mixture reaching the cylinders is rich in fuel (as rich as it's going to get).

The reason airplanes have such a thing as a mixture control (whereas automobiles don't) is that airplane engines are called upon to perform well over a wide range of altitudes. Since efficient combustion depends on the presence of both fuel and air in the proper ratio—and since air rapidly becomes thinner the higher one goes—it's easy to see that an airplane engine's fuel-air mixture *must* be adjustable if combustion is to occur normally at higher altitudes. (Otherwise, the mixture would become excessively rich, and power would be lost.)

There is a danger, however, in providing a manual mixture control on a high-output air-cooled engine. Aircraft engines require a certain amount of fuel (in excess of that needed for combustion) for sheer cooling. Reducing the fuel flow with the mixture control at high throttle settings removes some of this fuel cooling, while at the same time it increases combustion temperatures. The result is potentially damaging overheating of cylinders, pistons, valves, and exhaust components. (Also there's an increased likelihood of something called detonation—engine knock, in automotive parlance—which has been known to destroy engines. More of which in a later chapter.)

Obviously, the skill with which you manipulate your plane's mixture control can have a big effect not only on fuel economy, range, and endurance, but also cooling, reliability, and overhaul cost. The proper use of the mixture control will be discussed in Chapter 8. For now, it is only important that you remember that the mixture control (1) is purely mechanical, (2) adjusts the fuel-air ratio at the carburetor or fuel injector, and (3) operates independently of all other power controls.

Throttle

The throttle is often thought of as the primary power management control—the main ''go'' lever, so to speak. It pays to remember, however, that an aircraft engine's useful power output is actually determined by a *combination* of controls—mixture, prop pitch, turbo wastegate, carb heat (ram air in Mooneys)—of which the throttle is only one part. As a student pilot, it may have been convenient to think of the throttle as an ''accelerator,'' or main power control, but in flying more advanced aircraft you may find it conceptually useful (as I have) to think of the throttle as being the engine's main *air valve*—the control that modulates the flow of *air* into the engine. Follow the throttle cable from the instrument panel to the engine compartment, after all, and you'll see that in all cases—whether the engine is carbureted or injected, turbocharged or normally aspirated—the throttle is connected to an air valve (or a butterfly, as it's sometimes called). Push the throttle in, and you open the butterfly, permitting the engine to breathe more air. Close the throttle, and you close off the butterfly, making it harder for the engine to breathe. The carburetor or fuel injector, in turn, responds by doling out—or "metering"—fuel to the engine in roughly direct proportion to the volume flow of air going through it. But the main thing to remember is that when you move the cockpit control, you are moving an air baffle in the carburetor throat (or injector body); you are only indirectly controlling fuel flow. (Technically, in a continuous-flow fuel injection system, throttle movement produces movement not only of the butterfly but also of a fuel-metering cam or valve, thereby affecting fuel flow directly. Even so, the mixture control—not the throttle—has final authority over *fuel* flow.)

Perhaps it's worth adding that in most (but not all) aircraft carburetor installations, the butterfly shaft is mechanically linked, inside the carb, to a plunger device which acts to squirta single short blast of extra fuel into the carburetor throat when the throttle is rapidly opened. This plunger feature is known as an *accelerator pump,* and its purpose is to augment the carburetor's normal fuel output during transient periods of engine acceleration, so that the engine doesn't stumble or misfire while fuel flow catches up to air flow. The accelerator pump also enables the pilot to deliver raw fuel to the engine prior to starting (by stroking the throttle repeatedly), even though there's no airflow through the carburetor. This feature is not found in all aircraft carbs, however, and obviously in those aircraft that do not incorporate an accelerator pump (such as the Grumman Yankee/TR-2 series), throttle-pumping is not a useful technique for priming an engine.

Primer

While we're talking about fuel-delivery controls, we should probably say a word or two about primers. Priming systems exist in carbureted aircraft because the updraft positioning of the typical aircraft carburetor (along with gravity) ensure that any raw fuel that accumulates in the carburetor with the engine stopped simply falls to the pavement (or the bottom of the cowl, whichever comes first). Pumping the throttle sends fuel to the ground, not to the cylinders. In an automobile, of course, the carburetor generally sits directly on top of the engine, and gravity carries raw fuel straight down to the cylinders for starting. (Even so, a second air baffle—called a *choke*—is usually needed to enrichen the fuel mixture in an auto for starting.)

Primer systems vary in configuration from one plane to the next. Even within a given model series, you're likely to come across several types of primer setups. Some primers deliver fuel directly to the cylinders via individual fuel lines. In some systems, only one cylinder is fed fuel directly. Other systems are designed to deliver a spray of

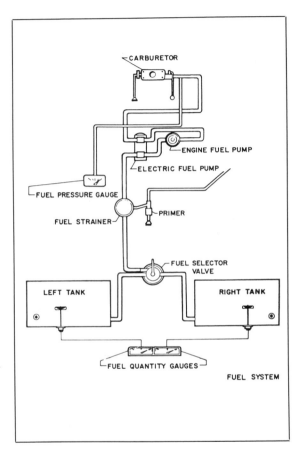

Fuel systems differ in layout from plane to plane, but the Piper Arrow schematic shown here is typical in having a primer that bypasses the fuel pumps, and a strainer downstream of the fuel selector. Electric fuel pump may either be parallel or in series with the motor-driven fuel pump.

fuel to a point (or points) in the intake manifold just downstream of the carburetor or injector body. Needless to say, it's a good idea to study the owner's manual to get an idea of what type of system your plane has. Better yet, pop the cowling and eyeball the installation yourself.

Some planes (mostly fuel-injected) do not have a primer system as such. Most fuel injectors used on modern airplane engines are themselves nothing but well-calibrated primer systems. Nevertheless, some injected airplanes do have separate primer systems. To some extent, this is because aviation gasoline is specially blended for ultra-low vapor pressure (maximum Reid vapor pressure: 7.0 psi) to forestall vapor lock at altitude. The low vapor pressure translates into low volatility and poor vaporization (especially in cold weather) for starting. This unfortunate characteristic has recently been offset—to some extent—by the increased use of aromatics in 100LL, and, of course, the legalization of autogas for aircraft.

The vast majority of primers now in use are of the manual piston/plunger (Kohler) type, wherein priming is accomplished by exercising a panel-mounted plunger (see illustration). The plunger works in syringe-like fashion to draw fuel from the main fuel strainer, or a point downstream of the fuel selector valve, up to the plunger mechanism itself, just behind the instrument panel; and from there, the fuel is sent directly to a discharge nozzle (or nozzles) in the intake manifold. Rubber O-rings on the end of the plunger provide a tight seal to keep raw gasoline from escaping into the cockpit, and also to keep fuel from creeping past the plunger (and to the engine) when the primer is not in use. Check valves ensure that vigorous plunger movement can send fuel in only one direciton: from the strainer (or main fuel line) to the intake manifold.

Note that the primer system's plumbing completely bypasses the airplane's main fuel delivery system, completely circumventing fuel pump(s), carburetor, etc. Thus, primer action is in no way dependent on the placement of mixture control, throttle, or indeed any other cockpit control except the fuel selector itself (should it be turned off).

Just so there will be no confusion: The proper way to use the Kohler-type primer is to unlock the primer by rotating the knob until its locking tangs clear the recesses in the knurled nut at the panel; pull straight out on the plunger until it is fully extended; wait two or three seconds to allow the plunger cavity to completely fill with fuel; then press hard on the plunger so as to create a forceful spray of fuel inside the intake manifold. (The idea is to atomize the fuel, rather than produce a small dribble of liquid.) After priming, push the knob all the way in, rotate it so as to lock it, and attempt to pull it back out (to see if it truly is locked). Failure to stow the primer in this fashion can result in fuel leaking past the plunger in flight, causing overrich

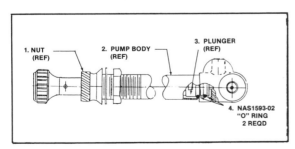

1. NUT (REF)   2. PUMP BODY (REF)   3. PLUNGER (REF)
4. NAS1593-02 "O" RING 2 REQD

The Kohler-type primer is nothing more than a finger-actuated piston or syringe sealed by rubber O-rings. The primer's output circumvents the carburetor, going straight to the intake manifold. For safety's sake, the primer must be locked after each use.

engine operation and possible power loss (to say nothing of unusually high fuel consumption, and trouble stopping the engine with the mixture control).

Prop Pitch

The propeller pitch control is an important power management tool (in airplanes that have one) because it controls engine rpm directly, more or less independently of the throttle, at all power settings above about 25 percent. Like the other primary engine controls, prop pitch is cable-operated: At one end of the Bowden cable is a control lever, or a vernier-type (infinitely rotatable) plunger; and at the other end is an engine-driven accessory known as a *prop governor*. Without getting too deeply into the technical details of prop governor operation, the governor is nothing more than a device for diverting engine oil, under controlled pressure, to the propeller hub, where—by exerting force on a movable piston attached to the blade pitch-change linkages—that oil can be made to do the work of changing the propeller's pitch. Inside the governor is a tiny oil pump (not much different, in fact, from the engine's own oil pump) which is capable of boosting incoming crankcase oil from around 50 psi to upwards of 200 psi. Also inside the governor are a set of flyweights that, in response to changes in engine rpm, control the opening and closing of a tiny escape hatch (technically known as the *pilot valve*—but again, let's not worry about technical details) through which that high-pressure oil can be selectively "leaked" to the prop dome. At the prop dome, the oil works against a piston to alter the pitch of the propeller blades. Excess oil is returned to the crankcase by means of a hollow passage in the crankshaft itself. In most systems, the high-pressure oil from the prop governor works against natural blade twisting forces to *increase* blade pitch (or decrease rpm). But in some installations—including most twins—the blades are preloaded to the high-pitch, or feathered, direction (with the aid of counterweights and springs), and prop-governor oil is used to *decrease* pitch (or increase rpm). In the latter case, a loss of engine oil pressure results in the prop automatically feathering. In the former case, loss of engine oil pressure sends the blades to the high-rpm position, all other things being equal.

If you've followed the discussion this far, you should now know why the manufacturers advise pilots to cycle the prop several times before takeoff on cold mornings (namely, because it takes several deep-cycles of the prop piston to displace cold, congealed oil in the prop dome with fresh, warm, "loosened up" oil from the engine). You also know why prop-governor leaks can be so messy (namely,

In Continental engines, the propeller governor usually can be found in front of cylinder number six on the forward crankcase (lower photo). Woodward governor shown above boosts oil pressure from 50 psi to 150 psi or more for prop actuation.

because the oil is under high pressure)—and why any oil leak from a constant-speed prop, or prop governor, is potentially hazardous (i.e., because the engine could suffer serious oil loss). But the main thing to know is that the prop pitch control simply acts to set the pilot-valve neutral point inside the governor to achieve the desired oil pressure inside the prop dome. Once this "setpoint pressure" is set, flyweights inside the governor take care of keeping the rpm right where it is.

Carburetor Heat

The carburetor heat control (usually a small push-pull knob grouped with the mixture and throttle) does just what the name implies: it heats the air going into the carburetor. The panel control connects, via cable, to a large flat-plate-type air baffle in the airbox at the entrance to the carburetor. With the cockpit control in the forward or "cold" position, the baffle sits at such an angle as to seal off the airbox to any incoming air except that which passes through the plane's engine air scoop and filter. When the cockpit control is pulled out, to the "hot" position, the giant flapper valve shifts to an entirely different angle to allow heated air to enter the airbox via a different routing. The heated air merely consists of unfiltered engine-compartment air that has been allowed to circulate through a shroud or heater muff placed around a section of exhaust pipe. From the shroud, the air travels through a large duct to the carburetor airbox.

The reason carburetor heat is necessary at all, of course, is that carburetors—by their very design—are susceptible to icing under quite ordinary atmospheric conditions. (Visible moisture is all it takes—i.e., if there's enough humidity in the air to produce visible mist, fog, haze, or rain, there's the potential for carb ice.) When air passes through the restricted area in the carburetor throat known as a *venturi*, a pressure drop takes place; and (in accordance with Boyle's gas laws) an accompanying temperature drop occurs as well. Also, when fuel is discharged into a high-velocity, low-pressure airstream, it evaporates—and cools—very dramatically. Combine the two effects together and you have the potential for potent refrigeration. In fact, a temperature drop across the venturi of 60 degrees is not uncommon. Which means that even with outside air temperatures as high as 90 degrees, it's still possible (if humidity is above approximately 60 percent) for airborne water vapor to "condense out" as ice crystals in the carburetor throat. When the accumulation of ice crystals reaches a sufficiently advanced stage, naturally, it interferes with the flow of air into the engine, and power is lost.

The insidious thing about carburetor ice is that it occurs gradually, and usually unbeknownst to the pilot. (Not many planes have carburetor temperature gauges.) By the time carburetor icing is noticed by the pilot—perhaps because the throttle is stuck and won't move any more (ice on the throttle butterfly can "freeze" the control solid)—the ice buildup often has reached such an advanced stage that a serious power interruption is unavoidable. As carb heat is applied, the venturi ice turns to water and is drawn into the cylinders, causing combustion to cease temporarily. (Combustion normally resumes once the water has passed.) But beyond that, the application of carburetor heat always brings with it *some* power loss, due to the greatly reduced density (and oxygen content) of the incoming heated air, and consequent overrichness. Lycoming has estimated the total power loss due to the density effect to be on the order of 13 percent. (With carburetor heat on, you're—in effect—flying under worst-worst-case density-altitude conditions: air temperature going into the engine is 130 degrees Fahrenheit!)

Three things to remember about the use of the carb heat control:

1. Tests by Lycoming indicate that carburetor heat can be applied continuously at power settings below 75 percent without ill effect on the engine. (Above 75 percent, there may be some tendency to hasten the onset of detonation.) However, if carb heat is to be applied continuously, the engine should be re-leaned to counteract the overrich condition created by the less-dense incoming air. Warm air is less dense (and contains fewer oxygen molecules) than cool air. Therefore, unless you are already operating on the lean side of peak EGT (see Chapter 8), the application of carburetor heat will bring on overrichness and make exhaust gas temperatures *drop.*

2. Unless your plane has a carburetor air temperature gauge and you are familiar with its operation, never apply *partial* carburetor heat under high-humidity conditions. Instead, leave the control all the way on, or all the way off. The use of partial carb heat can sometimes have the effect of warming a super-cold venturi (and incoming super-cooled ice crystals) to the point where ice will form *and stick* in the carburetor instead of passing through unnoticed.

3. Always remember that carburetor-heat air is *unfiltered* air, and unfiltered air is an invitation to premature overhaul. Periodically check to see that your carb heat control is indeed in the "off" position when not in use (some have a tendency to vibrate open in flight). Properly rigged, your panel control should have a few millimeters of "cushion" between the knob and the instrument panel when the heat control is off. That way, you know that the flapper valve is hit-

ting its stops before the panel control reaches *its* limit of travel, and not vice versa. (This is true for mixture, prop, and throttle controls as well, incidentally.) Have rigging adjusted if no cushion exists between knob and instrument panel.

Alternate Air

Fuel-injection systems are not as prone to induction icing as carburetors, since fuel evaporation takes place at the cylinders (where it's always warm) and since most injector systems lack a venturi. Nonetheless, under certain conditions, fuel-injected engines can fall prey to something known as *impact ice*, which is where ice crystals and/or supercooled water droplets (and/or snow) freeze in place after impacting the air scoop or filter. To deal with this eventuality, aircraft manufacturers provide a two-way baffle—actuated by a push-pull knob on the panel—in the induction system ahead of the fuel injector, to allow unfiltered engine-compartment air into the system should the main intake be blocked off in any way. Engine-compartment air is normally quite a bit warmer than ambient, of course; but alternate air is not normally heated, per se.

Operators of turbocharged aircraft have little to fear when it comes to induction icing or carburetor ice (note: some turbocharged planes, like the Cessna Turbo Skylane, are in fact carbureted rather than injected), since compressor discharge temperatures are normally high enough to preclude throttle icing. (This may not be true at low manifold pressures, however.) Nonetheless, impact ice can still occur. And here again, the answer is alternate air. But with many turbo installations, alternate air is *not* manually controllable from the cockpit; instead, the alternate air door is spring-loaded so as to open inward, automatically—via engine suction—when the airscoop or filter becomes blocked to the extent that the turbo compressor is sucking rather than compressing. Opening of the alternate air door may or may not be accompanied by some manifold-pressure loss.

Operation with the alternate air door open should be kept to a minimum, due to the fact that unfiltered air is being admitted to the engine. As with carburetor heat controls, the alternate air knob should not actually hit the panel when in the stowed position; some cushion should exist, to ensure that the "alt air" door is closing *before* the knob hits the instrument panel, not after.

Ram Air

A few aircraft—Mooneys, chiefly—have "ram air" knobs on their panel. In such aircraft, the ram air control is connected (via Bowden cable) to a trap door in the intake system which bypasses the main air filter and—when in use—permits prop blast and/or relative wind to enter the induction airscoop directly, thereby achieving (one hopes) some manifold-pressure increase. (An engineer might well explain the purpose of ram air induction as being to increase the *volumetric efficiency* of the engine in cruise operation.) That's the theory, anyway. In reality, the ram air control on, for example, a Mooney 201 can be counted on to produce no more than a fraction of an inch of manifold-pressure rise at 7,500 feet. The ram rise would tend to be greater at lower altitudes (for the same reason that indicated airspeeds are greater, with constant power settings, at lower altitudes), but in a plane like the Mooney 201, most cruising is done at 5,000 to 10,000 feet; and anyway, down low the ram effects would tend to be mitigated if, as is usually the case, the throttle is not firewalled. (The ram effect mainly takes place *ahead* of the throttle butterfly; any closing-off of the butterfly merely reduces the effect of the ram air.)

There are probably two lessons to be gotten here, if you're a Mooney pilot. First, if you want to achieve the maximum benefit (small as it is) from ram air, don't select ram air unless the throttle is open all the way; a partly closed throttle merely counteracts the ram effect. (Close the throttle all the way, and you'll counteract 100 percent of the ram rise.) Secondly, don't expect much ram effect on manifold pressure at high altitude, where your IAS (indicated airspeed) is low. An ideal altitude for use of ram air in a normally-aspirated airplane would seem to be 4,000 to 5,000 feet. Higher than that, ram pressure is low (due to thin air). Lower than that, you probably wouldn't want to use full throttle.

I suspect one reason the ram air feature doesn't produce much effect in the Mooney is that the normal induction system (filter and all) is already pretty efficient. The other reason it doesn't do much good is that the Mooneys are basically not "high indicated airspeed" airplanes at normal cruise altitudes. If you could indicate 250 knots in cruise, the ram effect would be substantial. (You can verify this, if you own a Mooney, by putting the aircraft in a shallow dive after selecting ram-air, and watching the manifold pressure build with indicated airspeed.)

A final reminder: Don't forget that ram air is *unfiltered air.* If you accidentally neglect to stow the ram-air knob on landing, you could end up ingesting quite a bit of abrasive silicon powder (dirt) on touchdown, particularly if you're operating out of a dirt strip.

Some ram effect, by the way, is present in many aircraft that lack a special ram-air door. If your manifold pressure increases in a dive—or decreases with alternate air—it's probably due to ram effects.

Cowl flaps affect not only CHT, but oil temp and accessory cooling.

Cowl Flaps

Cowl flaps are hinged panels or doors on the engine nacelle or cowl. Their purpose is to control the flow of cooling air out of (and into) the engine compartment during various flight regimes, so as to maintain acceptable cylinder head and oil temperatures. Most · high-performance aircraft have cowl flaps (although some, like the Aerostars and the Piper Malibu, do not). In single-engine installations—and some twins as well—cowl flaps are manually actuated, via a knob-and-cable arrangement. In larger aircraft, cowl flaps are electrically actuated (and when one of your cowl flap motors goes out, you're grounded until it's fixed, since cowl flaps have—in some twins—a critical effect on engine-out performance and engine cooling).

By opening or closing the cowl flaps, the pilot can increase or decrease the flow of cooling air over and through cylinder cooling fins—and also through the oil cooler (if one is present, which one usually is in aircraft that have cowl flaps). Thus, cowl flaps not only enable the pilot to optimize the cooling drag for any set of conditions, but give the pilot considerable control, also, over oil temperature—assuming that the engine's cooling baffles are in good shape and the cowling fits tightly. If baffling is missing, bent out of shape, or poorly maintained, cooling air will bypass cylinder fins, oil cooler, etc. and take the path of least resistance out the cowling—totally defeating the whole purpose of the cowl flaps.

We'll have more to say about the proper use of cowl flaps throughout this book. For now, it's enough to know that cowl flaps are designed to allow the pilot to vary the amount of cooling air that reaches the engine (not just cylinder cooling fins, but oil cooler, too), although not necessarily the accessory section, firewall, or rear cowl (it depends on the installation).

CHAPTER THREE
ENGINE INSTRUMENTS

Engine Instruments

As I talk to other pilots (new pilots, in particular), I am perpetually amazed at the vast amount of trust pilots seem to have in instruments which they have only a half-vast understanding of. Unfortunately, that trust is often misplaced. I prefer to see pilots gain a little more understanding, and leave trust on the ground where it belongs.

We will have a good deal to say about the inflight interpretation of various engine gauges in later parts of this book. The purpose of this chapter is to familiarize the reader with the principles of operation of the instruments themselves, so that one has a good idea of what might *really* be going on when a given instrument comes (or doesn't come) to life.

Tachometer

The tachometer is designed to display engine (crankshaft) rotational speed, usually in hundreds of rpms. In a plane with no manifold pressure gauge, the tachometer alone acts as the pilot's primary source of engine power output information under most flight conditions.

If you understand how a car's speedometer mechanism works, you're 90 percent of the way toward understanding how the mechanical tach in a single-engine airplane works. Basically, you have a long, flexible drive cable (enclosed in a flexible metal housing) going from a spot on the back of the engine near the oil pump or starter up to the instrument itself. At the gauge, the shaft connects to a cylindrical magnet that rotates inside an aluminum cup. The cup, in turn, is attached to the tachometer needle. As the magnet rotates inside the cup, eddy currents are produced which make the cup want to spin in the direction of the magnet; however, the cup is restrained by a spring, giving it only a certain amount of total movement. The actual amount of cup (and needle) movement is proportional to magnet—and thus engine—rpm. And that's about all there is to it.

Because of the routing distances involved, flexible drive cables are not practicable on most twin-engined aircraft, and for this reason, most larger airplanes employ electric (rather than mechanical) tachometers. The principle is simple: A tiny three-phase generator, mounted on the engine, sends alternating current to a synchronous motor in the instrument itself. The instrument motor, whose speed obviously varies directly with engine rpm, turns a magnet that sits inside a spring-loaded drag cup which, as above, is connected to the instrument needle. If the aircraft should experience an electrical outage, this type of tachometer will still operate, since it has its own AC power source.

Very few single-engine aircraft employ electric tachometers. One that does is the Mooney 231.

All aircraft tachometers have a silkscreened face, with a red line somewhere to indicate the maximum rpm that the engine is allowed to turn. (Operation beyond the red line, intentional or unintentional, usually means a costly engine teardown.) Most tachometers also have one or more green arcs, to indicate the normal operating range, or normal cruise range. Operation *outside* the green arc is allowed; but the green arc is there to indicate where recommended power settings are most often encountered. One or more yellow arcs are sometimes present on tachometers. They indicate cautionary zones in which transient operation is allowed, but not continuous operation. The most frequent reason for yellow arcs has to do with crankshaft-induced propeller bending modes.

The hourmeter on the face of your tachometer is nothing more than a tiny odometer mechanism geared to the rotating element with the tach. Different models of tachometer utilize different conversion factors (based on average expected engine rpm) to convert revs to hours. The tachometers used in 'C' through 'F' model Beech Bonanzas, for example (also Cessna 182s and most Mooneys) assume an average overall rpm of 2,310. Tachometers in later-model Bonanzas (plus Cessna 185, 206, 207, and 210 models) assume an average of 2,566 rpm. Throttle back, and you'll make the hours tick off more slowly; run wide open, and you'll reach TBO (recommended time between overhauls) quicker, simply because you are spinning the tach's hourmeter more rapidly.

The obvious shortcomings of the tachometer-type hourmeter (for purposes of accurate timekeeping) led to the widespread use of a different sort of "time elapsed" dial known as the *Hobbs meter*, which is separate from (and smaller than) the tach and likely to be found most anywhere on the panel. The Hobbs is a true electric clock (relying on the aircraft electrical system), coming on only when the aircraft is in use. The Hobbs may be wired to a pressure-sensing switch in the engine oil system, so that it only comes on when oil pressure rises above a certain threshold; or it may be wired to a microswitch on a landing gear leg (so that it comes on at takeoff, when weight is removed from the gear); or it may be wired in some even more novel fashion (to the airspeed indicator, for instance). Each plane is different.

Manifold Pressure

Manifold pressure is defined as the absolute pressure, in inches of mercury, of the air inside the engine's *intake manifold* (the system of pipes connecting the carburetor or fuel injector body to the cylinder intake ports). If you remember nothing else about manifold pressure, remember that it is always taken *downstream of the throttle butterfly*—never upstream.

The manifold pressure gauge itself is nothing more than an aneroid barometer whose sensing tube is plumbed into the engine air intake. In aircraft that have this gauge, it constitutes the primary power-reference instrument. The reason is simple: At any given rpm, engine power output is a function of the fuel and air flow through the engine (assuming the ignition is working properly, etc.), which is a function of throttle setting. Manifold pressure is a direct reflection of airflow into the engine. When the throttle is closed, impeding airflow through the carburetor, the pistons try in vain to suck air past the closed-off butterfly, dropping the manifold pressure to very low levels (perhaps just 10 or 12 inches of mercury, versus the normal ambient pressure of 29.92 inches at sea level). Conversely, when the throttle is wide-open, air passes freely into the carburetor as each piston descends on the intake stroke, and the pressure in the intake manifold remains close to ambient atmospheric pressure. Add a turbocharger compressor (to force extra air into the engine), and you can quite readily boost manifold pressure—and engine output—far above normal levels.

Because of restrictions in the intake system (a paper air filter, for example, plus the various twists and bends in the plumbing itself), the intake manifold never quite fills with air as fast as the pistons pump it away, even under wide-open-throttle conditions—which is why you'll always see at least a one-inch discrepancy between takeoff manifold pressure and ambient barometric pressure is an unsupercharged aircraft at takeoff. Let the engine come to a complete stop, however, and you should be able to read the correct ambient atmospheric pressure right off the manifold pressure gauge. (If not, you've got an inaccurate MP gauge, which will cause you to select inaccurate power settings unless you compensate for the gauge's error.)

Note that if you should ever suffer an altimeter failure during an IFR flight, you can use your manifold pressure gauge to get altitude information, since barometric pressure falls about one inch per thousand-foot increase in altitude. (For example, if your plane shows

28.5 inches of manifold pressure on takeoff at sea level, a full-throttle manifold-pressure indication of 26.0 inches in flight means you are at an altitude of approximately 2,000 feet MSL.)

Like all aneroid barometers, the manifold pressure gauge relies on a diaphragm (or pair of diaphragms), exposed on one side to high vacuum and exposed on the other side to an outside source (the engine manifold, in this case), to provide inputs to a moving needle. To prevent damage to the instrument in case of engine backfire, the pickoff line leading to the intake manifold usually incorporates some kind of restrictor, such as a coiled capillary tube or a restrictor orifice with needle valve. A variety of porous filter materials are also commonly used in the lines or fittings to discourage contamination of the instrument with fuel.

Manifold-pressure lines have been known to break, and when they do the symptoms are fairly predictable: The instrument indicates ambient pressure (30 inches at sea level) at all throttle settings, and the engine runs lean, perhaps stumbling or idling poorly. (The line break after all represents a sizable induction air leak.) Obviously, should such a condition be noted the aircraft should be grounded at once.

Oil Pressure

The oil pressure gauge is without a doubt one of the most important engine instruments in the cockpit, since it tells the pilots whether enough oil pump action is occurring to ensure adequate movement of oil through the engine (and thus continuous lubrication of moving parts). No oil pressure, no oil circulation; no oil circulation, no lubrication. No lubrication—no more engine.

The oil pressure gauge converts oil pressure to needle movement by means of a hollow, coiled metal tube known as a *Bourdon tube*. One end of this tube is open to the pressure source (an engine oil passage); the other end is capped off and linked to a meter movement. (Between the Bourdon tube and the engine is a long, stout piece of fuel/oil hose containing trapped air—or, in some cases, a volume of light oil or kerosene.) Under pressure, the Bourdon tube tends to uncoil and straighten out—like a child's party tooter—causing needle deflection. A model of simplicity, no?

Note that—as described—the oil pressure gauge is purely mechanical, involving no electrical devices. As long as the engine is turning over, this type of oil pressure gauge continues to function, even if your cockpit is blacked out by a total electrical failure.

Actually, there are some electric oil gauges in aviation; the Mooney 201, for example, uses a diaphragm-type pressure

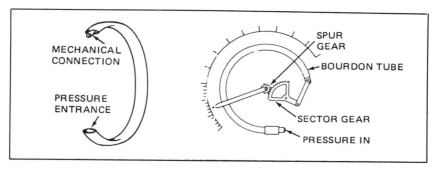

Bourdon tube mechanism.

transducer to convert oil pressure to a resistance change that can be displayed electrically on an ohmmeter-type gauge. In aircraft that use such a system, a sudden loss of all oil pressure on the gauge should be treated with suspicion unless accompanied by high oil temperature and/or sounds of grinding metal. Pressure transducers fail with much greater frequency than engines.

Take note, also, of the fact that under normal conditions, no engine oil ever actually reaches your cockpit gauge (any more than rain impacting your Pitot ends up inside your airspeed indicator). Should the Bourdon tube break, however, it is very possible for hot engine oil, under pressure, to come streaming into the cockpit behind the instrument panel. This only happens to one or two pilots a year, thankfully.

It's important to remember that the oil-pressure pickoff point—the spot at which the oil-pressure line connects to the engine—varies from installation to installation. In some cases, oil pressure is read immediately downstream of the oil pump, giving the pilot somewhat higher oil pressure indications than would be obtained otherwise. (This is true for Lycoming engines, for example.) In other engines, the oil must travel a considerable distance through and around the engine before arriving at the pickoff point, giving the pilot a somewhat delayed and (if anything) lower-than-normal indication of engine oil pressure. (This is true for most Continentals.) While the former setup responds more quickly to fluctuations brought on by oil-pump cavitation (or other causes), the latter approach is more apt to provide reliable cockpit indications in the event of upstream blockage of oil passages. A distal pickoff point allows the pilot to read *residual* oil pressure—what's left after oil has travelled to all or most of the key pressure-lube points in the engine.

We'll have more to say about oil pressure throughout this book.

Oil Temperature

Oil temperature may be measured in any of several ways, In some aircraft (almost all aircraft made before say, 1970), oil temperature is measured by coupling a Bourdon tube to a long capillary connected to a liquid-filled bulb immersed in engine oil. Pressure created in the bulb as the oil gets hot is transmitted via the capillary to the Bourdon tube and the instrument needle. (The pressure of the trapped vapor, obviously, is proportional to its temperature.) In other aircraft, a simple thermocouple is used. Like the vapor-pressure system just described, the thermocouple method has the advantage of not requiring any electricity from the plane's bus bar; the two dissimilar metals in the thermocouple provide *their own* electricity, which is roughly proportional to the temperature of the junction. The oil-temp gauge in a system employing a thermocouple is thus actually a tiny galvanometer, registering voltage on a scale marked in degrees Fahrenheit.

If you find that your oil temperature goes off-scale cold when you flip off your plane's master switch, you have still another type of oil-temp system—one which relies on the plane's electrical system to send current through a *thermistor probe* whose electrical resistance varies with temperature. Here again, the panel instrument is nothing more than a sensitive galvanometer, calibrated in degrees.

Regardless of the type of system installed, the oil-temperature sensor is usually situated at a point just downstream of the oil cooler (if one is present)—or near the oil pump, if there is no cooler.

Cylinder Head Temperature

Cylinder head temperature (CHT) means just what it says: CHT is the temperature of the cylinder head itself (the massive, finned end portion of the cylinder, containing the valves and rocker assemblies). The cockpit CHT gauge is usually calibrated in degrees Fahrenheit, with the green arc, or normal operating range, usually extending from about 250 to 460 degrees.

CHT measuring devices may be of either the thermocouple or thermistor type. (See discussion above, under Oil Temperature.) The latter type is now more popular. To determine which kind of installation you have, simply observe your CHT gauge at the end of a flight as you flip the master switch off. If the needle goes off-scale in the cold direction when power is cut, you've got a thermistor system. If, on the other hand, the CHT reading stays steady when you turn the master off, you've got a thermocouple system (or a stuck needle).

The key thing to note about your CHT gauge is that, unless an

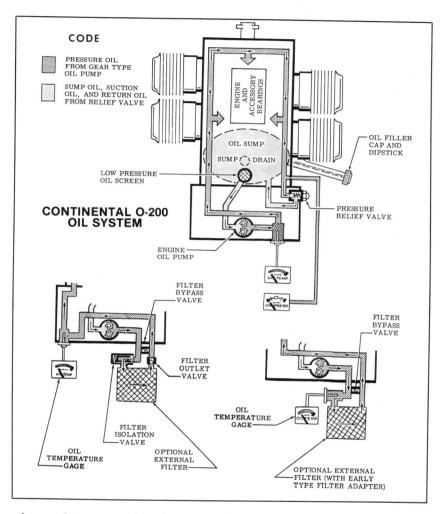

CODE

PRESSURE OIL
FROM GEAR TYPE
OIL PUMP

SUMP OIL, SUCTION
OIL, AND RETURN OIL
FROM RELIEF VALVE

ENGINE AND ACCESSORY BEARINGS

OIL SUMP

SUMP DRAIN

LOW PRESSURE
OIL SCREEN

**CONTINENTAL O-200
OIL SYSTEM**

ENGINE
OIL PUMP

OIL FILLER
CAP AND
DIPSTICK

PRESSURE
RELIEF VALVE

FILTER
BYPASS
VALVE

FILTER
BYPASS
VALVE

FILTER
OUTLET
VALVE

OIL
TEMPERATURE
GAGE

FILTER
ISOLATION
VALVE

OIL
TEMPERATURE
GAGE

OPTIONAL
EXTERNAL
FILTER

OPTIONAL EXTERNAL
FILTER (WITH EARLY
TYPE FILTER ADAPTER)

aftermarket system has been installed in the field, *the cylinder head temperature gauge is connected only to one probe, located on a single cylinder.* The airframe manufacturer determines (through flight test) which cylinder runs hottest, and locates the CHT probe there. From that point on, every bit of CHT information you get comes from that cylinder; the cockpit gauge tells you nothing about what's happening in the other cylinders.

This is a point worth underscoring. If you fly a 1980 Turbo Skylane RG with a single thermistor probe on cylinder number five, you have no way of knowing what the temperature of cylinder number one, two, three, four, or six is. Cylinders one through four might be in the latter stages of meltdown, and jug number six might be ready to come off, but you would never be able to tell it just by looking at the CHT gauge.

As said before, the manufacturers attempt to make the best of the situation by installing the CHT probe on the hottest (not necessarily the leanest) cylinder, as determined by flight testing; but even so, it pays to remember that with the sudden rupture of an exhaust gasket, an air intake leak, etc., *any* of your cylinders could instantly become the hottest cylinder—and unless your plane had special instrumentation, you'd never know it. If you can afford to, invest in a mutli-cylinder CHT system; it'll pay for itself someday in engine troubleshooting (not to mention inflight peace of mind).

Fuel Quantity
Your fuel quantity indicators (unless they're sight gauges) are electrically operated, drawing current from the aircraft battery and/or bus whenever the master switch is on. (The exact current draw is given in your aircraft service manual. Look for the load chart in the section on electrical systems.) The sensor at the fuel tank may be of two kinds. The most common kind—in planes as in cars—is the float-operated variable-resistance transmitter. Here, a plastic float bonded to a long pivot-arm bobs up and down on the surface of the fuel, varying the amount of electricity allowed to pass through the transmitter. Back in the cockpit, the ammeter-type fuel gauge ''sees'' these fluctuations in electricity and displays them as needle movements on an analog scale calibrated in pounds or gallons.

A second type of fuel-sensing system employs a large capacitor in each tank to measure fuel level; the varying amounts of fuel between the capacitor plates when the tank is full or empty cause the capacitance to change (due to the different dialectric constants of fuel and air), and this capacitance change is converted to needle movements on a fuel gauge. This type of system is found primarily (though not exclusively) on larger aircraft.

The key thing to keep in mind about your fuel gauges is that they are, by and large, pretty inaccurate—not because of any shortcomings of the mechanisms involved per se, but because aircraft fuel tanks are invariably wide and flat, and quantity is measured along the shortest dimension (height). Owing to the design of the tanks, it doesn't take much turbulence to set the fuel sloshing back and forth enough to make the fuel-quantity needles oscillate wildly (rendering the gauges

CHT and EGT readouts
can be digitized and
displayed on the same
gauge. This liquid-
crystal-display EGT-CHT
gauge by Electronics
International of
Hillsboro, OR is intended
for use on a twin-engined
aircraft.

virtually worthless). Worse still, the lower the quantity of fuel remaining, the sillier the gauges' behavior.

Don't ever take fuel gauges seriously. Floats hang up (or bonding
agent dissolves), rubber fuel cells pull up from the bottoms of
tanks (holding the float in the ''full'' position), wires cross,
resistances build up, float arms are bent in maintenance, etc. etc. In
my nearly 20 years of flying, I've found fuel gauges to be a valuable
crosscheck of other, more reliable sources of information on fuel remaining (such as a stopwatch and calculator, digital fuel totalizer,
past experience when flying a familiar route, etc.), but I have never
found it worthwhile to actually *trust* a fuel gauge of any sort.

Fuel Pressure

Fuel pressure gauges may either be of the Bourdon tybe type (see
discussion under Oil Pressure, above), or bellows-type. In the latter
type, a small bellows capsule—or two capsules—connects at one end
to a meter movement, and at the other end to a line leading to a point
downstream of the airplane's fuel pump(s). Pressure causes the
bellows to expand or contract like an accordion, giving rise to needle
fluctuations. As in the case of the oil-pressure gauge, no fluid actually
enters the instrument in normal operation. But there *is* a fuel line
coming to the cockpit, in this type system, and if a hose or fitting
breaks behind the panel, raw gasoline can flow into the cabin.

We will have more to say about fuel pressure at various points in this book. For now, suffice it to say that the fuel pressure gauge is useful for determining (before engine start) whether the auxiliary fuel pump is functioning, whether the engine-driven fuel pump is functioning (boost pump off), and—most important—whether vapor is forming in fuel lines.

Fuel Flow

The fuel flow gauge that comes as standard equipment on most fuel-injected aircraft is not really a flow gauge at all, but a pressure gauge calibrated in gallons per hour. Internally, there is no difference between this type of instrument and the bellows-type fuel pressure gauge described above. The only difference is where the gauge plumbs to the engine: The fuel-*flow* gauge connects to the fuel manifold or "flow divider" part of the fuel injector, which is the point where metered fuel— coming from the injector body, near the throttle valve—is evenly divided among the four or six (or eight) fuel lines leading to the individual cylinders. The pressure inside this spider valve (or flow divider) determines how forcefully the fuel will squirt out the injector nozzles into the cylinder intake ports. Or to put it another way, the pressure at this point is proportional to the rate of fuel flow to the engine.

The principal drawback to this arrangement is that when one of the injector nozzles becomes obstructed with foreign matter (as occasionally happens), increased back-pressure at the fuel manifold shows up at the fuel-flow gauge as an increased flow indication—when actually the flow rate hasn't changed. (It *can't* change, once the manifold pressure and mixture are set.) If half of an engine's injector nozzles are blocked off, the pressure across the remaining open nozzles will double, *and so will the indicated fuel flow.* To be sure, the fuel flowing through the remaining open nozzles has also doubled—but the *total* fuel flow to the engine has stayed the same. Hence, the fuel-flow indicator is (in this example) at odds with reality.

Bear in mind, then, that what you're looking at when you let your eyes fall on the fuel flow gauge is actually metered fuel pressure—not flow per se. Also bear in mind that regardless of the percent-power and altitude markings on the face of the dial, the fuel flow gauge is not (in and of itself) accurate enough to use for fine-leaning.

Fuel *totalizers* (or digitizers) of the kind sold on the aftermarket by people such as Insight, Silver, SDI (Hoskins), etc., often present fuel-flow information, but they use an entirely different sensing scheme than the original-equipment fuel-flow instrumentation described

Conventional analog-display EGT gauges come in switchable (left) and simultaneous-display (right) versions, the latter being more expensive but easier to interpret. (Alcor)

above. In a fuel totalizer system, there are no fuel lines going to the cockpit (a valuable safety feature of these systems over factory systems). Instead, fuel *flow* is measured directly—ahead of the firewall—by means of a flow transducer. The transducer is a flow-through device that uses a tiny pinwheel, the rotational speed of which is proportional to fuel flow. The pinwheel turns a serrated rotor which "chops" the light from a light-emitting diode; a photocell converts the light pulses to pulses of electricity, the pulses are counted electronically, and the tally is updated and displayed on a digital indicator on the panel.

The fuel flow information displayed by a digital fuel totalizer tends to be highly accurate (much more so than the indications given by a pressure-type fuel-flow gauge). But when the transducer's pinwheel hangs up, everything goes to pot. And unfortunately, transducer hangups are quite common in pinwheel-type systems. One manufacturer hopes to surmount this by means of a no-moving-parts transducer which uses ultrasound Doppler technology to measure fuel flow. Should that type of transducer become widespread, fuel-flow instrumentation will have taken a quantum leap forward in reliability and accuracy.

Exhaust Gas Temperature

The exhaust gas temperature (EGT) gauge constitutes one of the most versatile and powerful engine instruments on a modern panel. Without this borrowed jet-engine technology, operators of piston aircraft would still be in the Dark Ages where fuel economy and combustion troubleshooting are concerned.

The purpose of the EGT gauge, of course, is to tell you how hot your exhaust gases are. This has considerable significance for leaning (as we'll see in Chapter 8), as well as engine troubleshooting. For now, it's enough that you know, in general terms, what makes the system work.

The exhaust gas temperature gauge consists (in most cases) of a galvanometer needle moving on a scale calibrated in degrees Fahrenheit, connected by wires of carefully chosen length and diameter to one or more thermocouple probes mounted in one or more exhaust stacks. A system using one probe for each exhaust pipe (i.e., one probe per cylinder) is often referred to as an "exhaust analyzer." The cockpit gauge may display one cylinder's readout at a time, with a rotating switch to select cylinders, or it may present data for all cylinders simultaneously. Simultaneous-display indicators are available in analog (Alcor) as well as digital (Insight) formats. (Liquid-crystal single-cylinder displays are also available.) Note that while thermocouples are used, the probe output is generally amplified (and /or processed in other ways) before being displayed in the cockpit; and this requires bus voltage from the aircraft electrical system.

Note, incidentally, that the majority of EGT gauges installed in light aircraft today are calibrated in degrees Fahrenheit (25 degrees per scale gradation), but with no absolute temperature markings (and thus no redline). The lack of temperature numbers is not important in smaller aircraft, whose engines do not develop damaging exhaust temperatures (Cessna 150s, for example, hardly need to be placarded with an EGT redline), but in a turbocharged aircraft, or other high-performance plane, the absence of absolute EGT scales is a definite drawback, since damagingly high temperatures can be encountered in the course of normal leaning.

When getting checked out in a strange plane, it's often worthwhile to ask a few key questions about the EGT system. Is it a single-probe type, or a multi-probe type? If single, on which exhaust pipe is the probe located? (Or is it located in a 'Y' collector—in which case you're looking not at an individual cylinder's EGT, but the average of *several* cylinders.) Is there a gauge redline, and if so, at what temperature? Does the handbook allow operating at redline continuously, or only for brief periods while initially setting the mixture control? And: If there is only one EGT probe, is it on the same cylinder as the one that has the CHT probe?

We will return to the use of the EGT in later chapters.

Turbine Inlet Temperature

The TIT gauge is basically just an EGT hooked up to a single thermo-couple located at the turbocharger turbine inlet. It registers the temperature of the combined exhaust gases from all cylinders, just as they are about to pass through the turbocharger. In most cases the TIT gauge has an absolute temperature scale, with a redline at or near 1,650 Fahrenheit. It is extremely important (for reasons having to do with turbocharger life) not to exceed the TIT redline under any conditions.

Airplanes that have a TIT system may also have factory or after-market EGT systems (or exhaust analyzers) installed as well. The advantage of having both is that you can monitor combustion in individual cylinders as well as keep tabs on turbine inlet conditions. I personally favor having both, in a turbocharged plane. (The more information, the better.)

Compressor Discharge Temperature

Some turbocharged airplanes have a compressor discharge temperature (CDT) gauge in addition to TIT and/or EGT systems. The CDT works on the same principal, except that the thermocouple is placed in the induction system just downstream of the turbocharger compressor. For reasons having to do with detonation margins, CDT is usually redlined at or near 280 degrees F.

One parting comment regarding engine instruments, in general: Oftentimes, the engine gauges that require power from the airplane's bus bar share a common circuit breaker or fuse (frequently labelled "engine group"). In such systems, a short circuit in any *one* instrument in the cluster can, by popping the circuit break, render several others (fuel quantity, CHT, oil temperature) unusable as well.

Something else you should know is that Federal Aviation Regulation 91.33 requires that your tachometer, oil pressure, oil temperature, and fuel quantity gauges be in working condition prior to every flight. In addition, supercharged or turbocharged aircraft must have an operating manifold-pressure gauge. (Notice that CHT is *not* one of the required gauges.) If any of these instruments is not working, the aircraft, technically speaking, is grounded until repairs can be made.

CHAPTER FOUR
STARTING

Starting

Due to the myriad differences in control rigging, fuel metering systems, primer effectiveness, starter gear ratios, spark plug gaps, etc. from airplane to airplane, no single starting procedure—no simple mnemonic that I can give you—will suffice to explain how to start every engine. Installation differences (and ambient conditions) vary too much. Each engine is unique. Nonetheless, I would be remiss if I did not include in this book a chapter on engine starting. Over the years, I have too many times looked on with mixed feelings of empathy and horror as other pilots (and myself, occasionally) have ground their starter motors to nothingness, and drained their batteries to dead hulks, trying to get aircraft engines started under difficult circumstances. In most cases, such agony is avoidable.

In fairness to today's pilots, it should be pointed out that aircraft engines are not particularly easy to start by automotive standards, and have—if anything—gotten harder to start over the years, as engine displacement and compression ratio have increased (while batteries and starter motors have stayed the same). After all, the tiny four-cylinder engines that powered the Ercoupes and Aeroncas of 40 years ago were designed to be easily hand-proppable; the advent of electric starters made these low-compression engines almost as easy to start as an electric dishwasher. Since then, the engines have gotten bigger—and balkier. Considering the design factors at work, perhaps it's no wonder "hard starts" are so common. Today's Lycomings and Continentals are behemoths compared to modern car engines and present starter motors with tremendously high torque demands. (One reason for the high friction drag: aircraft lubricating oils are several grades thicker than the oil you put in your car.) Yet airplane batteries are smaller and have *less* cold-crank reserve than car batteries.

Then there's the fuel. Aviation gasolines are purposely blended to have a low vapor pressure compared to auto fuel—one result of which is that avgas doesn't vaporize as readily as car gas. Also, aircraft fuel delivery systems are generally of the updraft variety, which means most of the (raw) fuel needed for starting wants to dribble out of the engine by gravity before reaching the intake ports. (In most cars, by contrast, the carburetor sits on top of the engine, and fuel has nowhere to go but straight down into the intake manifold.) Airplane owners don't even get the benefit of an automatic choke.

To top it all off, aircraft magnetos deliver a pathetically weak spark for starting compared to automotive electronic ignition systems. The voltage output of a magneto increases with engine rpm, which means that at cranking speeds, mag output is at its lowest. Thus, aircraft spark plugs are limited to relatively narrow gaps (.018-inch nominal). Automotive solid-state ignitions, by contrast, are designed to deliver a terrific spark of custom-tailored waveform at *all* engine speeds. (Plug gaps of .050-inch are quite acceptable in most cars, as a result.)

All things considered, perhaps it's remarkable that pilots can start their engines at all—particularly given the brief and oversimplified starting instructions in most owner's manuals. (Engine starting is one area in which cookbook airmanship of the type that results from over-reliance on preprinted checklists will almost certainly let you down, sooner or later.) Reason enough to go into some detail here with regards to the mechanics involved.

General Principles

Always remember this: An internal combustion engine *must* fire when air, fuel, and a spark are present in the combustion chamber. When an engine won't start, it is because there is too much air, too much (or too little) fuel, and/or too little spark—or too *early* a spark. The only other things that can keep an engine from starting are mechanical problems of fairly major proportions (things like stuck valves, mis-timed camshaft, holed pistons, etc.) which, thankfully, are rare.

While there is no such thing as a Universal Engine Starting Checklist, all engine-start procedures have in common the fact that they are designed to bring air, fuel, and spark to the cylinders in the right proportions (and at the right time) for combustion to occur. A typical start sequence might go as follows (this checklist was actually taken from the Continental O-470 Operator's Handbook):

1. Carburetor heat: OFF.
2. Propeller pitch: FULL FORWARD (high rpm).
3. Cowl flaps: OPEN.
4. Fuel selector: FULLEST TANK.
5. Throttle: CRACKED (open approx. 1/10 of total travel).
6. Mixture: FULL RICH.
7. Radios and switches: OFF.
8. Master switch: ON.

9. Boost pump: ON.
10. Primer: ENERGIZE.
11. Ignition: "BOTH."
12. Starter: ENGAGE.

All engine-start checklists (for carbureted engines, at least) are a variation on this theme. Going down the list item by item, you can see that there's a clear logic to the list, and its order. Carburetor heat is best left in the off position, because you don't have carburetor ice, and you don't want unfiltered, hot air going into the engine during startup. Prop pitch is left in the high-rpm position because that's the position (generally speaking) that puts the least oil-pressure demand on the engine lubrication system. (Also, you want the engine to pick up speed quickly after it fires, and to do that you need low propeller pitch.) Selecting *low rpm* during startup may cause the engine to run erratically and die after combustion occurs.

Cowl flaps are best left open during starting because air-cooled cylinders need all the cooling air they can get during ground operations. (Notice how fast the CHT comes up after starting an engine.) The fuel selector should be on a tank that has fuel, for obvious reasons.

Next, we come to cracking the throttle. In aviation, cracking the throttle does not mean bending the throttle handle back and forth until it fatigue-fractures. It means advancing it just off the fully closed position (i.e., opening it a half inch or so). There really is no "best" way to crack the throttle. Most pilots have a certain respect for the throttle which causes them not to "crack" it open far enough. (A quarter inch is not far enough.) On the other hand, you don't want to open the throttle, say, halfway. During starting, particularly in cold weather, you want the throttle butterfly to act as a manual choke, blocking off air and artificially enrichening the mixture. It can't do this if you open it more than a quarter of the way. Hence "cracking" the throttle.

The mixture control, you'll recall from Chapter 2, is your master fuel on/off valve. With the mixture in idle cutoff, no fuel gets to the engine (no matter how far open the throttle is); for starting, you want a good deal of fuel to get to the engine. So put it "full rich" for starting.

Radios and switches are left off for starting for the simple reason that very damaging voltage surges are often sent through the aircraft electrical system during engine startup. (Also, line voltage drops very low during cranking, as the starter motor draws up to 300 amps of current.) These fluctuations can severely damage solid-state devices in your radios. Also, you want the full output of your battery to be available for cranking. Therefore, leave your radios, strobes, autopilot, courtesy lights, and other unnecessary electrical accessories off for starting. (That includes the "alt field" portion of your master switch, if present.)

The master switch can be turned on any time in the start sequence prior to engaging the starter. (There is no particular reason for making "master on" the eighth item on the checklist.) The point is, you do need to have your battery on-line before the starter motor will come on. Many novice pilots have found this out the hard way (i.e., by complaining to the FBO mechanic that the starter is dead, when in fact the master switch was never turned on—an embarrassing confrontation). Turning the master on also allows your boost pump to work.

Although many handbooks are silent on the subject, I prefer to leave the "alt" portion of a split master switch off during starting, and I recommend you do the same. There is no reason to want to have the alternator field energized at this point—all it does is heat up your field windings and drain amperage from your battery that could better be put to use cranking the engine. Also, there's always the chance (if the alternator is on) that an electrical transient could damage your alternator diodes during engine cranking. Remember, your alternator doesn't come on-line until the engine is already running at pretty good speed (1,200 rpm or so); it's not needed for starting. Leave it off.

The boost pump is turned on later in the list to conserve electrical energy, and to keep the pump itself from wearing out. Strictly speaking, it's not needed for starting a carbureted engine unless there's vapor or air in the fuel lines. It's a good idea to turn it on before start-up, however, not only to purge vapor but to check to see that the pump itself (and your fuel-pressure gauge) works.

Now we come to the only real "discretionary" item in the checklist (the only item calling for any real judgment on the part of the pilot)—namely, priming. This is where it pays to know and understand your fuel-delivery system in some detail. If your plane is equipped with a primer system, learn what kind of system it is and how many cylinders are under its control. If your plane is carburetor-equipped, you'll want to know what kind of carburetor you have (float or pressure), and whether there is an accelerator pump in it. Float carburetors with accelerator pumps can be used for priming (as can pressure carburetors); but generally speaking, if there's a primer system installed, it's best to use the primer to prime, and leave the throttle alone. The primer delivers fuel to the cylinders more directly; and in any case, due to the carburetor's updraft positioning, pumping the throttle only causes raw fuel to go to the pavement (or the bottom

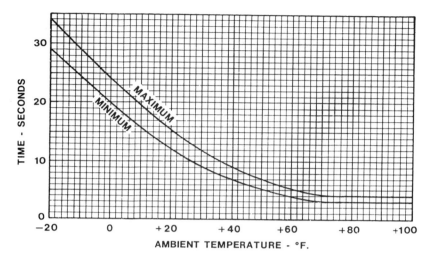

This chart, from the Piper Turbo Arrow pilot's operating handbook, tells the pilot exactly how long to prime in cold weather. Unfortunately, most POHs lack such a chart.

of the cowling—a fire hazard in either case). Also, throttle-pumping is a very imprecise means of metering fuel. Besides which, if the plane is fuel-injected—or if your carburetor lacks an accelerator pump— jockeying the throttle will have no effect whatsoever.

If you're not accustomed to using your primer to start your engine (your instructor taught you to pump your throttle), get in the habit of using it. Follow the instructions in your owner's manual. The Cessna 182L manual, for example, advises using two to six strokes of the primer plunger under normal conditions (more if ambient conditions are cold; less if warm). Do remember to fully close and lock the primer knob when you're done.

If the engine is fuel-injected, you'll use a slightly different checklist than the one shown above. The continuous-flow port fuel injection systems used on light planes are essentially nothing more than finely calibrated full-time primer systems. To prime for starting, you generally just run all the power knobs (mixture, throttle, prop) to the full-forward position, activate the electric fuel pump for five to ten seconds or until a positive indication is seen on the fuel-flow dial, then retard the throttle about three quarters of the way. Presto —you're primed.

After priming, it's time to turn the ignition switch to "both" (unless the owner's manual advises something else), then—finally

—engage the starter. Generally speaking, if you've done a fair job of estimating the amount of prime needed, the engine will fire within two crankshaft revolutions, then pick up speed. As soon as the engine fires, you should release the starter button or key (to prevent possible damage to the starter—and also to cut out the retard breaker action, if your mags are of the "shower of sparks" type); then watch the oil pressure gauge. Don't fiddle with the radios, talk to your passengers, or engage in any other activities until oil pressure has come up. *Keep your eye on the oil pressure gauge until the needle comes off the peg.* Lycomings will generally show a pressure indication sooner than Continentals, and in cold weather a longer wait for oil pressure is to be expected, since cold oil flows slowly. But in any case, if no pressure indication is seen within 30 seconds, *shut the engine down at once.* Serious damage will occur if the engine is run without oil pressure.

Once the engine has coughed to life, don't forget to turn the alternator-field portion of your master switch on before powering-up your avionics.

Fuel-Injected Engines

There is nothing inherently different or mysterious about starting a fuel injected engine. The objective, as always, is to get air, fuel, and spark to the cylinders. Various handbooks offer various techniques for achieving this; some say to place the mixture in the full rich position, *then* turn the boost pump on, *then* open the throttle until a fuel-flow indication occurs, and turn the boost pump off (and retard the throttle) before hitting the starter. Others say to put *all* power levers full-forward, *then* hit the boost pump, *then* pull the throttle back for starting. Some owner's manuals advise cranking with the mixture in idle cutoff, advancing it only after the engine comes to life. My strong advice is to try the handbook procedure exactly as written, before listening to anyone else's advice. If the handbook method fails to give good results, develop your own starting procedure (gleaning what tips you can from other operators, and perhaps from this book). Only experience will teach you what's right for your engine.

Injected Lycoming engines (IO-360, IO-540, etc.) come with Bendix fuel injection systems as standard, whereas Continental's injected powerplants (IO-470, IO-520, TSIO-520, etc.) have Teledyne Continental Motors (TCM) fuel injection systems. The TCM and Bendix systems are functionally very similar: both provide continuous injection of fuel to the individual cylinder intake ports, *outside* the combustion chamber. (Since the intake valve is open only about a third of the time when the engine is running, this means that raw fuel is sprayed against a *closed* intake valve two-thirds of the time. Not efficient, but it works.) The Continental system meters fuel in proportion to engine rpm via a rotary fuel pump. The Bendix system, by contrast, meters fuel in proportion to airflow, as sensed by a venturi with impact tubes in the throttle body. The Bendix injector's airflow-sensing apparatus is not particularly sensitive to low airflows, and thus for starting purposes meters fuel rather haphazardly at normal cranking rpms, whereas the TCM rotary fuel pump delivers gasoline in a much more controlled fashion during cranking. For this reason, TCM systems generally give more predictable starting characteristics, in the author's opinion. (Lycoming engines tend to fire up predictably with adequate priming, but keeping them running after letting go of the starter button is often tricky under hot-start conditions.)

Neither the Bendix nor the TCM type of fuel injection can deliver fuel to the engine for priming without the use of an auxiliary fuel pump. (Throttle-pumping does absolutely nothing to prime a fuel-injected engine.) Likewise, even with the boost pump on, neither system will allow fuel to pass to the cylinders with the mixture control in idle cutoff. Nor will you get much fuel flow with the throttle closed. *The throttle, mixture, and boost pump must be used in concert to prime an injected engine.* The same priming considerations that apply to carbureted engines (see above) also apply to injected engines: namely, prime little or not at all if the engine has been run within the last 15 minutes; and prime more in cold weather than in warm.

Flooded Start
When an engine produces a fews short bursts of black smoke on a start attempt, then dies and won't come back to life, it's because the engine is flooded (or overprimed). Paradoxically perhaps, *too much* fuel can be just as much an impediment to starting as *too little* fuel. Fortunately, there is an easy cure.

To get an engine unflooded, you need to greatly increase the amount of air it's receiving, and purge any accumulated raw fuel from the system, if you can. Getting rid of accumulated fuel is often best done by having a cup of coffee while excess gasoline drips from the induction system drain tube. (This is also a good way to confirm that you have a flooded start. Get out of the cockpit and kneel down next to the cowling. If you can see—or smell—raw fuel anywhere, chances are good you're flooded.) In any case, the important point to remember is: the engine is now primed, and all you have to do is adjust the fuel/air ratio to give more air and less fuel, and the engine will start. Set the mixture control to the idle cutoff position (thereby

Injector plumbing crosses directly over hot cylinder fins, allowing fuel to boil out of delivery lines after shutdown. Hot starts require careful use of mixture, boost pump.

preventing any more fuel from getting to the cylinders); open the throttle all the way (to allow air into the engine); and crank with the ignition on "both." Within several revolutions, the ratio of fuel to air will decrease to the point where combustion can occur. As the engine fires, you should simultaneously retard the throttle and advance the mixture. The engine will then continue to run normally.

Inexperienced pilots often approach this procedure with some trepidation, fearing perhaps that (with the throttle wide open) the engine will somehow "get away from them" when it roars to life. Nothing could be further from the truth. With the mixture in idle cutoff, the engine is incapable of running for more than a second or two after it "catches," even with the throttle full-in. There is no danger of the engine accelerating to redline (unless, of course, you advance the mixture to full rich and forget to retard the throttle). Set the parking brake, in any case.

The foregoing technique is applicable not only to fuel-injected engines, but carbureted ones as well. The mixture-lean part of the procedure keeps the engine from getting more fuel; the throttle-open part assures that the cylinders get massive amounts of air. (The carb's accelerator pump will squirt a small amount of extra fuel into the engine—or onto the ground, as the case may be—at the moment you open the throttle, but the amount is negligible compared to the quantity of air that will follow.) Again, when the engine fires, simply advance the mixture and retard the throttle.

If you're unsure whether you've flooded your (injected) engine on a balked start, by the way, you can always proceed to flood it intentionally, then go through the "flooded start" checklist just outlined. (Don't laugh. I know one Certified Flight Instructor whose standard Lycoming IO-360 start technique—cold, hot, or otherwise—is to flood the engine, then use the "mixture-lean, throttle-open" technique until it starts. "It's more reliable than the book procedure," the man explains.) To *flood* an injected engine, push all the power knobs in, hit the boost pump for ten seconds, and pull the mixture back to idle cutoff. (Turn the pump off.) Do not, under any circumstances, leave the boost pump on longer than ten seconds, as you may create a *hydrostatic lock* condition in the cylinders (which is where raw fuel accumulates in a pool inside one or more combustion chambers). Hydrostatic lock is a sure invitation to cylinder-head loss, rod breakage, crankcase cracking, and/or other mischief.

Hot Starts
Hot starts (any start attempt which begins with a hot engine—regardless of ambient conditions) are probably responsible for more extended-cranking episodes, especially with fuel-injected engines, than any other type of start. Fuel injection poses a particular problem because the individual 1/8-inch stainless-steel fuel delivery lines to the cylinders sit immediately above the hot engine, absorbing heat by radiation as well as convection. Usually, the small amount of fuel trapped in these lines after a hot shutdown develops vapor bubbles prior to the next start attempt (if indeed it doesn't boil away entirely). In addition, fuel inside the injector body (behind or under the engine) may turn to vapor. This vapor must be cleared before the engine will start again.

Generally speaking, if a hot engine has been shut down no longer than 10 or 15 minutes, the pilot's initial hot-start procedure should be simply to crack the throttle and crank the engine without further priming. The shorter the shutdown time, the better the success of this technique. In any case, though, your first strategy in restarting a just-shut-down engine should always be to crank without priming (throttle closed). If the engine is carbureted rather than injected, try a small amount of throttle priming (one or two partial strokes before cranking). Vary the amount of prime in accordance with the amount of time the engine has cooled off.

If the engine has sat for more than 15 minutes (but not long enough to be cool to the touch), you'll have to use to a vapor-clearing technique before restarting. Here, the trick is to purge the vapor from the engine without flooding it in the process. (As always, try the procedure given in your airplane owner's manual or Pilot's Operating Handbook first, before resorting to other techniques.) Under hot-day conditions, when you can be sure that the fuel in various parts of the system ahead of the firewall has accumulated vapor, begin by opening your throttle halfway, retarding the mixture to *idle cutoff,* and energizing the electric fuel pump for 30 seconds. This procedure causes relatively cool fuel from the main tanks to be sent to the injector throttle body, where it displaces warm (possibly boiling) gasoline and sends it back—via the return lines—to the tanks. Warm fuel is thus replaced by cool fuel, and the heat-soaked injector body is stabilized at a lower temperature.

Next, advance the mixture to full rich, turn the ignition to "both," activate the boost pump (use the "low" position, if the switch has one), and wait for a positive fuel-flow indication to appear on the flow gauge, signifying the arrival of fresh fuel in the spider valve (or flow divider) atop the engine. Depending on how much vapor is present, up to 60 seconds may elapse before a stable fuel-flow indication is seen. When the fuel-flow needle peaks, turn the pump off and begin cranking. As the engine catches, reduce the throttle to a fast-idle setting. *Be ready for vapor reappearance.* If there was vapor in the spider valve or individual injector lines, it will flow to the injector nozzles within a second or two after the engine coughs to life, and the engine will die just as quickly as it started. Don't let this happen. When and if rpm starts to fall off, momentarily activate the boost pump again, to purge the residual vapor. (Rock the boost pump switch to the high position intermittently, if necessary, to clear the vapor, but don't keep the pump on any longer than necessary or you could flood the engine.)

Throughout the above procedures, remember that during a hot start, you're beginning with an already-hot starter motor, and extended cranking will only overheat it. Never crank for more than 30 seconds continuously without allowing a three-minute cooldown period. Heat can easily damage the armature.

Also, once again, if oil pressure doesn't come up within 30 seconds, shut the engine down and investigate for leaks or other signs of distress.

If at any time you suspect you've flooded the engine (with a hot engine, the line between priming and flooding is a very thin one indeed), go immediately to the "Flooded Start" techniques on p. 63.

Cold Starts

In the absence of preheat and power carts, cold-weather starting (i.e., OAT below freezing) can pose a significant challenge. Aviation lubricating oils—extremely thick by automotive standards—are positively taffy-like at subfreezing temperatures, greatly increasing the friction drag for starting; yet at the same time, cold OATs reduce the available output of wet-cell batteries. (I might add that both effects—the effect of cold on oil viscosity and the effect on battery output—are exponential, not straight-line.) At some sufficiently low temperature, the cranking power required exceeds the power available, and you're "behind the power curve" for starting. At that point, only a jump start—possibly combined with preheat—will get you going.

The arrival in the late 1970s of multiviscosity oils for aircraft has done a great deal to help pilots out of the cold-start dilemma (affording engines extra protection from cold-start damage in the process). The cranking power required when an oil like Aeroshell 15W-50 is used can be less than half that required when SAE 40 (or even SAE 30-weight) oil is used. But multigrade oils are no substitute for preheat when outside air temperatures (OATs) go below 20 degrees F. The manufacturers recommend preheat at temperatures below 20 degrees, whether or not you're using multivis oil, for the simple reason that engine tolerances are so tight at low temperatures, it's extremely easy to encounter cylinder scoring, piston scuffing, and/or ring breakage during a cold-start attempt, regardless of how good the oil is. (Remember that cylinder barrels are tapered towards the head end, so that at operating temperature the barrel contour will be cylindrical. Before CHT comes up, the rings are being rammed into a very tight space at the top of piston travel. At low temperatures, the cylinder metal contracts, and narrow clearances become even narrower.) Again: Below about 20 degrees, you should count on using external pre-heat before attempting to start your engine, regardless of what type oil you have in the sump.

When pre-heating an engine, by the way—whether you're using a hair dryer or a propane combustion heater—be sure to alternate between the top of the engine and the bottom (i.e., between directing hot air at the cylinders, and at the oil cooler and sump). A good practice is to force hot air into the front of the cowling for five to ten minutes, then force air up through the cowl flap(s) for five to ten minutes, then go back to the front of the cowling, etc. You want the cylinders *and* the oil in the sump to be warm before you start, and that means alternating the heat from top to bottom. (It also means spending more than just 10 or 15 minutes heating the engine. Face it: No 450-pound mass of metal is going to absorb a significant number of BTUs in less than 20 or 30 minutes' time.) Do not, however, direct hot air at heat-sensitive plastic parts, nor at fuel injectors, carburetor bowls, vacuum pumps, or other accessories containing heat-sensitive elastomers.

Incidentally, pre-heat your battery, too, if you have time. (I used to actually remove my Skylane's battery and bring it home between winter sorties. It made cranking go much faster on cold mornings.)

When pre-heat has been used, prime your engine as if standard-day conditions applied. Remember that with a pre-heated intake system, fuel is going to vaporize readily and combustion will occur with only moderate priming.

If pre-heat is not available, the main problem in getting your engine to start (assuming the battery can provide sufficient juice to crank it over) will be getting enough *fuel vapor* to the cylinders for combustion to occur. This generally means lavish priming. Read your owner's manual. If the manual says to use six strokes of the primer, use six strokes—don't be shy about it. With experience, I found that getting a Skylane's Continental O-470-R to start in freezing weather was not difficult, even with a slow-turning starter, so long as I gave the engine six strokes of primer, *turning the prop through one revolution by hand between each stroke.* Turning the prop through serves several functions. First, it loosens congealed oil. Secondly, it allows time for the raw fuel to evaporate. Thirdly (and very importantly), it cycles all intake valves through the open position so that all cylinders can admit fuel vapor. Fourthly, it redistributes oil on the cylinder walls, preventing the raw-fuel washdown that would otherwise occur. (Caution: Observe proper safety procedures whenever you turn a propeller by hand. Be sure the switch is off, and have a safety pilot in the cockpit. Treat the prop as if the ignition is hot—because it just might be, if you have a broken P-lead.)

My Skylane cold-weather starting regimen also included a couple more procedures. One was to leave the primer plunger out while cranking and press it in as needed to keep the engine running once combustion occurred. Another thing I learned to do to keep the engine running was to apply full carburetor heat for starting (at temperatures below 40 degrees). Exhaust gases being fairly hot even at idle, the carb heat—it turns out—is effective almost as soon as the engine comes to life. Carb heat serves two purposes during a cold start: First, it aids vaporization of the fuel, improving combustion and evening-up the distribution of gasoline to the various cylinders. (Poor distribution is a hallmark of large carbureted engines.) Secondly, by reducing the density of the incoming air, it artificially enrichens the fuel/air mixture beyond what can be obtained with the mixture control in full rich. A little extra richness never hurts, particularly considering the fact that at freezing temperatures, the air is extra-dense (hence extra-fat with oxygen—making your mixture artificially *lean* unless you do something about it).

At one time, the manufacturers offered fuel-dilution kits as options for cold-weather operators. Such kits allowed pilots to pull a knob just before engine shutdown at the end of a day's flying, sending raw gasoline into the engine oil system. After preleasing a specified amount of gasoline, the engine was shut down. On the next startup, the fuel-diluted oil (naturally being less viscous) would allow vigorous cranking and rapid flow of lubricant through the engine. By the time the engine came up to normal operating temperature, most of the fuel would have evaporated off, and the cycle could be repeated at the end of the flight.

Some manufacturers still allow oil dilution; Cessna, in some of its manuals, even proposes (for operators without dilution kits) draining oil between flights and combining it with a specified amount of gasoline before putting it back in the sump for a cold-start. In general, though, oil dilution is now conceded to be a bad idea (except as a last resort), certainly inferior to the use of multiviscosity oil and external pre-heating, for a variety of reasons, not the least of which is that if you haven't used oil-dilution in a season or two, you're apt to dislodge a season's worth of sludge from internal oil passages. (The sludge can then travel like blood clots to distant parts of the engine, causing havoc.)

Automotive gasoline, because of its generally higher vapor pressure, ought to give slightly easier cold-starting than avgas, from better vaporization. (Indeed, the whole reason auto fuel is blended to have high vapor pressure in the winter is to aid starting.) This is hardly a reason, by itself, to use auto fuel, however. If auto fuel is to be used, it should be clean and of high quality (high octane and not adulterated with methanol); and the engine and airplane should have Supplemental Type Certificate approval for auto fuel use.

One hears of various and sundry schemes to preheat aircraft engines. Virtually everything that's been done to cars has also been tried with airplanes, including electric dipsticks, battery warmers, cowl blankets, light bulbs in the cowling (probably too much a fire hazard to be considered worthwhile), electric blow-dryers, propane heaters, etc. Pilots in the far north are accustomed to using very lightweight oils, and even preheating the oil (in a separate container) before each flight—a technique that might not make much sense with multigrade oil, since the oil would thicken as it heats up.

One of the best preheat systems (for those owners who have access to 110-volt AC electricity) is the Tanis method. In this system, electric heating elements are screwed into pre-existing CHT-probe bosses in the cylinder heads, and elsewhere, so as to provide well-distributed preheat to cold-sensitive parts of the engine. (The probes are hard-wired to the engine via a harness; the owner simply plugs a cord into a 110-v AC outlet to activate the system, preferably the night before a given flight.) The Tanis system is lightweight, reliable, safe (no fire hazard), and relatively inexpensive at $400 to $800 installed. But you do have to have access to 100-volt electricity. (Or a very large grid of batteries.)

Some owners have experimented with piping the hot exhaust from their car's engine straight into the aircraft engine compartment (via heater-duct hose, or ''skat'' tubing). This approach leaves much to be desired, however, since car exhaust is loaded with ozone (harmful to rubber hoses in the airplane's engine compartment), sulfur dioxide (very corrosive), soot, unburned gasoline, and oil vapor, not to mention steam (which can condense out on a cold engine), carbon monoxide, and other impurities. Almost any method of preheat is preferable to the car-exhaust approach.

Obviously, there is no substitute for a heated hangar. (If you can borrow the use of one for 30 minutes before each flight, so much the better.) Forced-air heat from an electric heater or propane combustion heater is also effective, as is the Tanis system. And it goes without saying that if cold-weather flying is a regular part of your life, multigrade oil should be considered mandatory. (Battery preheat is a good idea too.) The main thing to remember is that when the outside air temperature dips below 20 degrees Fahrenheit or so, each startup (in the absence of *some* form of preheat) brings with it a definite risk of cylinder scoring, piston scuffing, tappet spalling, and—in general—extreme engine wear. Plan accordingly.

Jump Starts

Except for Mooney 201s and 231s, all piston-powered planes in current production use 24-volt, negatively grounded wet-cell batteries, and thus cannot be jump-started from anything other than a 24-volt power cart (which, thankfully, most large FBOs have). On the other hand, Mooneys and light aircraft made before about 1976 generally have 12-volt systems, and *can* be jump-started with an American-made car or any source of 12-volt DC power. If your plane has an APU (auxiliary power unit) receptacle, you can jump-start it with the appropriate power cables (pilot's mail-order shops sell both common types for about $50); otherwise, you can hook regular automotive jumper cables to the aircraft battery, following the standard negative-ground, plus-to-plus procedure.

Be sure to read your aircraft owner's manual before undertaking a jump start. Depending on how your plane's starter motor is wired, you'll want to have the master switch on or off before hooking up the jumper cables (the owner's manual will tell you how to proceed). In the absence of other information, try leaving the master switch off (including the ''alt'' portion of a split rocker) for the jump, and have the car's operator rev the engine before engaging the airplane's starter. *Be sure to observe proper polarity (negative ground, positive terminal to positive terminal) when connecting jumper cables, and be certain all electrical accessories—radios in particular—are off before proceeding. Also be sure the plane's brakes are working and any assistants are well briefed on safety precautions to be observed in the vicinity of whirling propellers.*

It should go without saying that a jump start—while it may get your engine going in extremely cold weather—is *not* a substitute for preheat at OATs below 20 or 25 degrees.

Due to the placement of sensing probes, Lycoming engines generally develop oil pressure more quickly upon startup than Continental engines. During a cold-weather start, oil is slower to flow and some delay can be expected in oil-pressure indications. Accordingly, Lycoming operators should allow up to 25 seconds for an oil-pressure indication to show up after a cold-start; Continental owners can let up to 45 seconds go by in extremely cold weather. If these time limits are exceeded, stop the engine at once and do not attempt a restart until thorough preheating has been accomplished.

Hung Starts

What happens if you can't get the engine started (or more often happens, the engine coughs to life and abruptly quits)? Always check out the simple things first. In cold weather, the most likely cause of an abortive start is underpriming. In hot weather, vapor lock and overpriming are the usual cuplrits. When the engine turns over but will not fire at all (not even the slightest cough of life), and there's no smell of raw gasoline (indicating flooding), the first thing to suspect is ignition. Are the mags on? Is the ignition switch working? These things sound idiotically simple, but remember, where there's fuel, air, and a spark, there's *always* combustion. You can't introduce gasoline, oxygen, and a spark to a combustion chamber and not get an exothermic reaction.

In very cold weather, when an engine fires up and then dies after 10 or 20 revolutions (and then won't restart to save your life), the cause is very often *ice bridging* of the spark plug electrodes. This occurs because normal combustion produces very large quantities of water vapor (and carbon dioxide), and the water condenses on ice-cold electrodes to produce droplets of ice. Since the ice is non-conductive, the spark plugs are thus rendered inactive. Starting cannot be accomplished until the ice melts and the plugs are dry. (You can remove the plugs from the engine and dry them individually, or aim a hair-dryer at each one *in situ* for about five minutes.) A good preheat will, of course, prevent ice bridging.

Mixture distribution to the various cylinders of an aircraft engine is rarely perfect at normal operating rpms, and at idle (or cranking) rpm, the fuel/air distribution is even worse than normal. Thus it often happens that when an engine is slightly overprimed, or under-primed, two or three cylinders (sometimes just one cylinder) will fire on startup, with the rest sputtering or doing nothing, in which case the engine lopes along at a pathetic 200 or 300 rpm after you let go of the starter button, two jugs firing and the rest starving for fuel (or air). The best remedy for this situation is usually to *slowly* and smoothly advance the throttle (mixture rich, boost pump off) to about an inch open. The cylinders that are getting the proper mixture will generally fire more energetically and the engine will pick up speed—and the other cylinders will, often as not, benefit from the rpm change. (Fuel distribution is better at higher rpms.) Quickly advancing the throttle is not recommended, since it will usually result in a backfire and complete engine stoppage. If you suspect that the non-firing cylinders are not getting enough fuel, energize the boost pump or Kohler primer (as applicable) briefly and see what happens. Often, this will get the balky jugs to light. (If, on the other hand, you suspect you may have overprimed the engine, try leaving the throttle where it is and slowly leaning the mixture. Aircraft carburetors are routinely set overrich for idling, and the slightest amount of overpriming can lead to the "hung start" syndrome described above; momentarily leaning the mixture will often restore rpm quickly and get combustion going in all cylinders.)

Backfiring

If the engine backfires (particularly in cold weather, when a generous amount of prime has been used), be alert to the possibility of an engine fire. I make it a practice to carry a fire extinguisher in cold weather, and I recommend you do, too. The standard advice in case of a carburetor fire is to keep on cranking the engine, in hopes of sucking the fire into the combustion chambers, thereby containing it; but experience shows that most pilots who have had carburetor fires were not aware, at the time, that a fire occurred (only later is the air filter found to be charred and the carburetor discovered to be fire-gutted). If you have an engine fire and can see it from the cockpit, it's time to get out of the airplane. Turn the fuel off, grab the extinguisher, and go. Don't continue cranking.

Backfiring deserves special attention, because it's almost always a tipoff to improper starting (or priming) technique, or something seriously wrong with the aircraft ignition system. A faint popping sound during a hard start (assuming the propeller keeps turning the proper direction) is nothing to be concerned about—but a loud backfire (especially if accompanied by a prop kickback) should definitely get your attention. It means you should take a time-out to consider the possible causes—and to investigate for damage (to the starter ring gear, induction piping, and air filter).

A kickback almost always signals trouble with impulse couplings or retard breaker action. If you're not familiar with the terms "impulse coupling" and "retard breaker," these terms refer to special features of aircraft magnetos designed to aid starting. It would be impossible to start a modern-day aircraft engine without either impulse couplings or retard breakers.

Consider how a magneto works. A magneto is nothing but a rotating magnet in a circuit employing some coils of wire (to step up the voltage created by magnet rotation) and a mechanically tripped switch called a *breaker* (or "points," in automotive parlance). The mag develops a whopping big spark—up to 10,000 or 20,000 volts—by virtue of inductive flyback action, as magnet polarity reversal sets up *flux changes* that produce voltages that are stepped up by the coils. The important thing to notice is that magnetos—by their very design—develop more electrical power as rpm increases. This is both a strength and a weakness. At high engine rpm, you want a good, strong spark. But at normal engine cranking rpms, magneto output is so low as to be worthless. The "coming in" speed of many aircraft mags (the speed at which usable spark is produced) is 200 to 300 rpm—well above the cranking capacity of most starters.

And then there is the problem of spark *timing*. Even if a magneto *could* deliver a powerful spark at cranking rpm, the ignition timing of modern aircraft engines (at 20 to 30 degrees before top center of piston travel) is such that, without special provisions for delaying (or retarding) the spark, an engine being turned over at, say, 100 rpm might well "kick back" on firing—damaging the starter motor and/or other components.

One answer to this problem—the problem of how to get a magneto to deliver a *strong, late* spark for starting—is a device known as an impulse coupling. The impulse coupling is a spring-and-flyweight affair interposed between magneto and engine in such a way that—during cranking—the coupling winds up, in clockspring fashion, before "snapping" the magneto rotor (i.e., the magnet itself) through the firing position rapidly, and late. The device's design is such that as soon as the engine starts up, the impulse feature disengages automatically through centrifugal flyweight action. At all rpms above one or two hundred, the impulse coupling simply acts as a drive adaptor, and the magneto doesn't know it's there.

Most Slick magnetos come with impulse couplings. Quite a few Bendix mags do, too. You can tell if your plane has impulse-equipped mags by hand-turning the propeller in the normal direction of rotation and listening for the loud metallic *clank* of the coupling(s) as you pass top dead center of each piston's compression stroke. (Caution: Treat the prop as if it were "hot," even though the ignition is off.) One or both of your mags may have impulse couplings.

As ingenious and reliable as impulse couplings are, they sometimes give trouble. For instance, the flyweights can become magnetized, locking in the "closed" position so that they do not engage their stop-pins during low-speed cranking. Or, in cold weather, oil can congeal on the flyweight pivots, temporarily immobilizing them in the "off" position—rendering the magneto useless for starting.

Even when impulse couplings are working properly, however, they leave something to be desired in that they produce only one brief spark on each piston stroke—which may not be enough to initiate and sustain combustion under severe cold-weather conditions. Under many conditions, it would be much more desirable to have a *long-duration* spark at each plug when the piston reaches top dead center on its travel. (But of course, this spark would have to occur late enough in the cycle to preclude engine kick-back.) The "shower of sparks" ignition system pioneered by Bendix does just this. It provides a slow-cranking engine with an energetic spark, late in the combustion cycle, to aid starting—without some of the drawbacks of the impulse coupling.

In the shower-of-sparks system, a small device called an *induction*

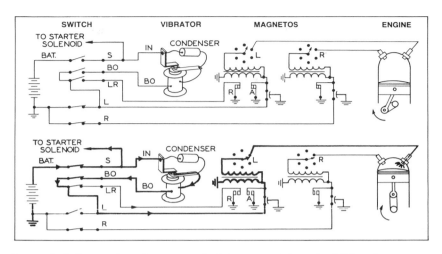

In a typical shower-of-sparks system, the left mag contains retard (R) and advance (A) breakers. When starter is engaged, pulsating battery current is sent from a vibrator to the primary winding of the left mag via the P-lead. When the retard breaker opens, a "shower of sparks" is created at the spark plug.

vibrator is used in conjunction with an extra set of breaker points inside the magneto (known as the *retard breaker*) to create a late, hot spark for starting. The vibrator—an electrical accessory that draws power straight from the aircraft battery—comes on when you engage the aircraft starter. Its function is to send a continuous stream of pulses (around 200 pulses per second) of electricity to the primary winding of the magneto so that when the retard breaker points open (very late on the compression stroke), a more-or-less nonstop stream of sparks is generated at the spark plug—a "shower of sparks." This sparking lasts as long as the retard points are open—i.e., through about ten degrees of crankshaft revolution (in contrast with the single, quick spark that jumps the gap once, and only once, when an impulse coupling trips). When an impulse coupling fires, the spark lasts only a few milliseconds—it is extinguished long before the mag's regular breaker points close. Not so the "shower of sparks" system.

The purpose of the retard breaker points in the shower-of-sparks magneto, as mentioned above, is to ensure *late* delivery of the spark for starting. These points (which are driven off the same cam as the mag's normal, or advance, breaker points) are wired through the ignition switch in such a way as to render them inactive any time the starter is not engaged. (Ditto the induction vibrator.)

Obviously, when dealing with kickback problems on a balked start, it helps to know whether you've got "shower of sparks" or impulse-coupled magnetos. With a "shower of sparks" system, a kickback could mean a wiring problem in the ignition switch, or in the mags themselves. In an impulse-type mag, a kickback almost always signals lack of impulse action. Either situation deserves attention.

With any normal magneto having just one set of points, breakage of the primary lead (the so-called P-lead) results in a permanently "hot" mag—always "on" because it cannot be grounded through the ignition switch. A "shower of sparks" mag, however, has (in essence) *two* "hot wires,"—one to each set of points—*both* of which must be functional for the mag to work right. If the wire to the *advance* breaker is open or broken, or has a poor connection at the mag, you'll not only have a "hot" mag (and zero rpm drop on runup), but there will be no way for vibrator current to get into the mag to create a shower of sparks. (The vibrator current arrives at the mag's primary coil by way of the advance-breaker lead.) If the wire to the *retard breaker* should break (or the connection is bad), you'll get vibrator action—because the vibrator will be grounded through the advance breaker—but the retard action will be lost, and chances of a kickback are excellent.

The moral should be obvious: Check magneto connections (both "advance" and "retard"), plus breaker timing, whenever starting difficulty is experienced with a "shower of sparks" type of magneto.

And with *any* magneto, check internal timing (E-gap) as well as mag-to-engine timing, should kickback be experienced persistently.

Hand-Propping

If your battery is dead and you're absolutely unable to secure a jump-start, you may be tempted to hand-prop the engine. Be advised, however, that if your magnetos are of the "shower of sparks type," some battery power is necessary to excite the induction vibrator. (Without battery power to the vibrator, you'll get no shower of sparks—and no start.) Also, in a shower-of-sparks system, the ignition switch *must* be held in the "starter engaged" position for starting (even when hand-propping), in order for the retard-breaker circuit to work. Failure to activate the retard breaker will result in a nasty kickback on starting—a highly dangerous state of affairs, if you are hand-propping.

In general, hand-propping an aircraft engine is a bad idea, unless the engine is of small displacement (200 cubic inches or less) and low compression ratio (no more than 7.0-to-1). Propping a Lycoming IO-540 or Continental IO-520 should be unthinkable. (Certainly your insurance company will tell you as much.) Only hand-prop an engine if it is the last resort; if the engine is small; if there is an experienced pilot in the cockpit; the brakes work; and your confidence level is high. Do not hand-prop an engine otherwise.

Morning Sickness

If your engine starts okay, but idles very roughly for the first few minutes, smoothing out only after a thorough warmup, you are witnessing the telltale first sign of sticking valves. This syndrome (rough startup followed by smoothing out as the engine comes up to normal operating temperature) has been referred to by some mechanics as "morning sickness." It occurs most often with low-horsepower Lycoming engines, and most often on the first start of the day.

When an engine starts roughly and idles roughly for the first few minutes, shut it down and look under the cowling. If you don't know what your pushrod tubes look like, have a mechanic point them out to you. (Lycoming pushrods are located above the cylinders; Continental pushrods are below the cylinders, parallel to the direction of piston travel.) Look for bent or bowed pushrod housings. Any deviation from perfect straightness signals pushrod bending (and valve sticking), and should mean automatic grounding of the aircraft.

Valves sometimes stick in mid-flight (giving rise to a rough engine in midair), but more often than not, if a valve sticks, it's usually an exhaust valve, and it usually will stick just after a hot shutdown, when engine parts are contracting. Deposits on the valve stem will freeze the valve in place, tending to hold it partway open (typically), so that on your next start attempt, it doesn't want to move. There is precious little "give" in the valve train, however, so if the valve fails to travel, something has to bend. Usually it's the pushrod.

To preclude engine damage due to "cold stuck" valves, always pull the engine through before the first start of the day. A stuck-closed valve will reveal itself as a hung prop (a prop that's hard to budge); a stuck-*open* valve will reveal itself as a total lack of compression—lack of a compression "hump"—during pull-through on one cylinder. In either case, have the situation checked out by a mechanic before further flight.

CHAPTER FIVE
RUNUP

The Pre-Takeoff Runup

The pre-takeoff runup is one of the most misunderstood procedures in all of flying. Virtually every pilot is taught how to go through the motions of conducting a runup (or "mag check"), starting, usually, with one's very first flying lesson; but almost nobody really understands what to look for in a runup, or—more important—how to interpret a bad one. For most pilots, it is a missed opportunity—a (failed) chance to get to know the airplane and engine a little better.

Contrary to what some pilots seem to believe, the purpose of the pre-takeoff runup is not to try out the ignition switch and see if the tachometer is working. The purpose of the runup, simply put, is to determine if the plane's engine is in a condition to fly. (To some extent, runups are performed also to see that the carburetor heat works, and—if the plane is equipped with a constant-speed propeller—to ascertain whether the prop governor is functioning. But mainly, the runup is done to judge the airworthiness of the powerplant.)

In this chapter, we'll go through the standard runup procedure (and its components) in some detail, then discuss the interpretation of "trouble signals," so that the reader can make intelligent go/no-go decisions while perhaps also isolating faults (and saving the repair technician some troubleshooting time).

The Basic Runup

Handbooks vary quite a bit on what constitutes an acceptable runup rpm. Some, like the Cessna 150A manual, say to run up at 1,600 rpm; others, like the Piper Cherokee 235 handbook, advise much higher rpms, such as 2,150. Which is as it should be. There is nothing magical about the choice of rpm for the runup; it varies from plane to plane and is largely arbitrary to begin with. By far the most important thing is that you settle on a particular rpm value for any given plane, and be consistent from runup to runup, so that long-term trends can be monitored. The actual rpm you choose is of little practical consequence, as long as the engine is developing useful power and you aren't operating in a yellow arc on the tachometer.

The present-day predilection for 1,800- and 2,000-rpm runups is to some extent a carry-over from past tradition. In years gone by, when radial engines (and tailwheel landing gear) were the vogue, pre-takeoff runups were customarily done at full throttle. Engines were slow-turning (many were redlined at 2,200 rpm or less), and prop clearance being what it is on a conventional-gear airplane (namely, tremendous), there was little reason *not* to run up at full throttle, especially considering that engine reliability was then, much more than now, often suspect. The best way to know whether an engine was ready for takeoff, circa 1940, was simply to run the thing up to takeoff rpm for a brief check. If the engine was smooth and all ignition systems checked out, a safe takeoff was virtually assured.

Today, we still do our runups at 2,000 rpm or so, not because there is anything sacred about that particular rpm, but simply because it is impractical to run up at full throttle. Redline rpms are much higher on today's aircraft engines than on pre-Korean-War designs, and prop tips operate much closer to the ground (a mere inches away, in fact) on tricycle-gear aircraft. Ground-running a modern light plane's engine to redline only accelerates damage to prop blades from sand and gravel, sends the same gravel to nose-gear

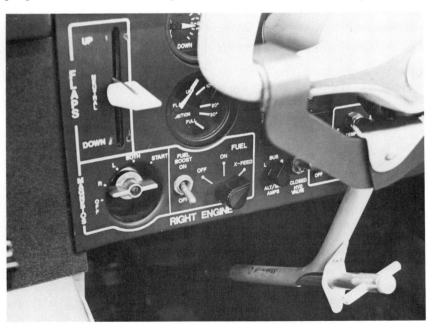

Twin-engined airplanes may use toggles or a rotary selector switch (as in this Aerostar) for ignition.

oleos and wheel fairings (where it does additional damage), stirs up dust which can get into the engine (particularly during the carb-heat check), and buffets tail surfaces mercilessly without benefit of aerodynamic damping. Clearly, full-throttle runups are no longer appropriate. Modern aircraft engines are reliable enough that their airworthiness can as often as not be adequately determined by a brief run to 50 percent power (i.e., 2,000 rpm or so). I hasten to add, however, that a full-throttle runup should by no means be ruled out as a followup to a faulty low-rpm (standard) mag check.

It goes without saying that if you have the handbook to the plane you intend to fly, you should adhere to the specific runup procedure contained in it, before trying any ad-lib technique. Nevertheless, the day will no doubt come (if you fly often enough) when you'll have a chance to fly an unfamiliar plane without benefit of an owner's manual or decent checkout, in which case you should have *some* kind of makeshift runup procedure to fall back on. The following runup procedure (taken from the Cessna P210N Pilot's Operating Handbook) is typical of the cookbook procedures given in most factory manuals. With common-sense modifications, it can be adapted for use in almost any piston-powered airplane.

ENGINE RUNUP:

1. Cowl flaps—OPEN.
2. Parking brake—SET.
3. Mixture—RICH.
4. Throttle—1,700 RPM.
5. Magnetos—CHECK INDIVIDUALLY (rpm drop should not exceed 150 rpm on each, or 50 rpm differential between mags).
6. Propeller—CYCLE.
7. Engine instruments and ammeter(s)—CHECK.
8. Suction gauge—CHECK.
9. Throttle—REDUCE TO IDLE.

If the airplane in question has carburetor heat, the carb heat knob should be tried somewhere between steps 5 and 9. A significant rpm loss—at least 200 rpm—should occur when the heat is turned on, as the less-dense incoming hot air throws the effective fuel-air ratio off in the rich direction. If little or no rpm loss occurs, it may be because the heat valve is faulty—in other words, you may without knowing it be running with fulltime carburetor heat. This would be a dangerous condition from several standpoints, not the least of which is that the engine would not develop full power on takeoff. Note that a correctly rigged carburetor heat control achieves full turning-off of the heat before the knob contacts its ''cold'' stop. (If the knob actually *hits* the stop, you have no way of knowing that the heat valve, at the carburetor, is fully closed. Only if the valve closes before the knob hits the stop can you be sure.) Note also that when operating with carb heat on, unfiltered air is used for induction and rapid engine wear can occur (particularly inasmuch as you're stirring up dust by running the engine up on the ground). Keep carb-heat operations to a minimum during ground operation.

On airplanes that do not have carburetor heat, there is often something called an ''alternate air'' control. If you see a knob or lever so labelled, exercise it during the runup between steps 5 and 9, and check to see that the engine operates normally with alternate air selected (some small manifold pressure or rpm drop is acceptable). Cessna omits this step from the P210N checklist above because the P210N induction system uses an automatic alternate air door, the operation of which is beyond the pilot's control. (There is no ''alt air'' lever in the P210N.) This, of course, will vary from plane to plane, even within model runs. Early Mooney 231s, for example, have manual alternate-air controls, while later 231s have an automatic system. Automatic systems should receive periodic inspections by a qualified mechanic, since they cannot be checked on pre-takeoff runup.

Air-cooled engines get very hot very fast when operating on the ground (watch your cylinder head temperature, and you'll see), which is, of course, why the checklist begins with the item ''cowl flaps—OPEN.'' Accordingly, ground runups should be carried out in as expeditious a manner as possible. As a rough rule of thumb, the entire procedure outlined above should not take more than 30 seconds, for a single-engine aircraft. Also, when possible the runup should be performed with the aircraft *facing into the wind.* (If no wind is present, hold in position directly behind the aircraft in front of you, and use his prop blast to cool your engine.)

Whenever performing a pre-takeoff runup (which I prefer not to call a ''mag check,'' since checking the mags is in fact only one item on the list), a definite order should be followed: Brakes on, *then* throttle opened to 1,700 rpm (or whatever), *then* the magneto check, *then* cycle the prop and look at the power (and suction) gauges, *then* throttle down again. Why check the mags before cycling the prop? If it's a cold morning, you'll give the oil more time to warm up—and anyway, if one or the other ignition system isn't functioning properly, you're not going to be flying and you won't need to cycle the prop. (Cycling the prop is fairly traumatic for the engine—much more so than any other phase of the pre-takeoff check—and you might as well not do it until you know the mags work.) Why check the engine instruments before reducing rpm to idle? Because you want to know if the oil pressure, oil temperature, and cylinder head temperature are indicating properly with the engine putting out power. (What the gauges say at idle is of less importance.) Why check the ammeter before reducing throttle? Because the alternator or generator can't reliably be checked at idle rpm—the "coming in" speed of most generating systems is around 1,200 rpm. Similarly, your vacuum pump is designed to produce usable suction at *flight rpms* (1,500 and up), not at idle—so check it while the engine is at runup rpm. Takeoff rollout is no time to discover that your gyro instruments are dead.

A word of caution: The foregoing "engine runup" checklist is in no way meant to be a complete *pre-takeoff* checklist; depending on the aircraft, there may be quite a number of other things to attend to before deeming all systems "go" for takeoff. (For example, you should check the fuel selector for proper positioning, see that the flight controls are free, set your directional gyro, check the cabin door lock, and carry out numerous other checks before considering the airplane ready to fly.) In the event that you find yourself in an unfamiliar plane with no checklist, the best pre-takeoff heuristic, I find, is not to try to remember somebody's nonsensical acronym or mnemonic, but simply reach out and touch every instrument, every switch, and every control that you can touch with your fingers; and as you touch each one, look at it and ask yourself what its function is, and what it should indicate for takeoff. Start with your seat belt; then see that the cabin door is secure. (Is there a door lock? What position should the lever be in?) Is there an autopilot? Touch it to be sure the switch is off. Is there a circuit breaker panel on the ceiling? Touch the rows of breakers to see that none is popped. Touch the trim handle(s); the fuel selector(s); all engine controls; all radios (set the frequencies); all switches; all instruments (set the DG and altimeter, and wind the clock); and then sit back and look to see if there's anything you haven't touched. The "Braille checklist" routine seldom fails, no matter how complicated the aircraft, *if* you take the time to work your way around the entire cockpit. I recommend it highly—but only as a last resort, when no factory checklist for the airplane can be found.

Some pilots prefer an abbreviated runup on the second (and third) flight of the day. I can't recommend it, though, unless you are solo, the plane is not complex, and your insurance is paid up.

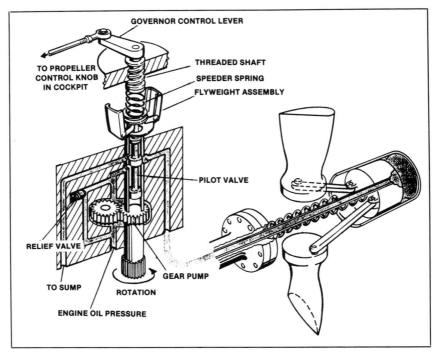

The prop governor has its own gear-type oil pump which boosts crankcase oil from 50 or 60 psi to well over 150 psi. Pressure for prop actuation is "leaked" in a controlled fashion by the pilot valve (which opens or closes in response to flyweight action and/or pilot control inputs). Changes in rpm are "felt" by centrifugal flyweights which, as they pivot, open or close the pilot valve as needed to maintain constant rpm selected by the pilot.

Propeller Function

A function check of the propeller governor system is an essential part of the pre-takeoff runup on any airplane that is equipped with a controllable-pitch propeller. Nowadays, "controllable pitch"—for all practical purposes—means constant-speed, as there are very few bonafide manually controllable props in service any more. (Those of you with electric Beech propellers on your Bonanzas, skip to the next section.)

A surprising number of pilots—particularly those who do most of their flying in fixed-pitch-propeller planes—forget (or never knew) that the hydraulic action in a constant-speed propeller system depends on the availability of oil pressure from the engine oil system.

Prop oil is engine oil. The prop governor (a gear-driven accessory at the front of a Continental engine, or at the rear—typically—of a Lycoming engine) is little more than an oil pump, drawing from engine oil galleries, which boosts oil pressure to the values necessary for prop actuation, with the exact amount of pressure determined by a valve controlled by centrifugal flyweights. Engine oil is thus sent from the governor to the prop dome, where it acts against a piston (which in turn connects to pitch links at the prop blades) to effect rpm changes. Return oil goes from the prop dome to a hollow passage in the crankshaft, ultimately to the engine oil sump.

The proper way to check a constant-speed prop during a runup is simply to pull the prop pitch control all the way aft (into the "feather" position, if it's a featherable prop) and *hold it there* until the tachometer begins to register a sharp decrease in engine rpm. How sharp is sharp? On a warm day, or with warm oil, 200 rpm is plenty—anything more than that needlessly stresses the prop and crankshaft (and attached components). On a *cold* day, with cold oil, you should wait until the rpm decays at least 400 rpm. And regardless of what the oil temperature gauge is telling you, *always* cycle the prop more than once on a cold day, before takeoff. The oil temperature probe, remember, is near the oil cooler—it says nothing about the temperature of the oil inside the prop dome. In the prop dome, you've got last week's oil, sitting there (not circulating), probably in a congealed state, perhaps mixed with a little air. The idea is to clear this old oil out of the prop dome, and fill the hub with fresh, warm, air-bubble-free oil. To do this requires that you deep-cycle the prop several times—maybe as many as five or six times.

On a cold morning, don't be surprised if up to 20 or 30 seconds elapse before the propeller responds to the cockpit rpm control. (Hint: Multigrade oil flows much better at low OATs than straight-weight oil, which means you'll get better prop actuation. Do switch to multigrade oil if you intend to fly where the weather is nippy.) Air trapped on the "oil" side of the prop actuating piston can have a delaying effect, too, as can sludge buildup inside the crankshaft. (The oil-return passage in the crank tends to collect soft carbon sludge rapidly, due to the centrifuge effect of the crank's rotation. This passage should be cleaned every 200 hours in planes using straight mineral oil, or in planes that are flown infrequently and receive infrequent oil changes.) If the prop doesn't respond at all to the cockpit control, needless to say, you should taxi back in to the maintenance hangar and have the situation corrected. *Do not* take off with a sluggish prop in hopes that further engine operation (in flight) will warm the oil to the point where normal prop actuation occurs. In the air, the prop dome tends to stay very cold (colder than almost any firewall-forward point on the airplane) due to blast-cooling, and congealed oil is bound to stay that way for the duration of the mission. Get your prop looked at if it doesn't respond well on runup.

Engine Instrument Indications

An important part of the pre-takeoff runup is noting engine-instrument indications. (This check should be repeated, if possible, on the takeoff rollout.) All engine gauges should be in the green before takeoff (with perhaps one exception—see next paragraph). The oil pressure must be above the bottom of the green, but less than redline. Likewise, oil temperature should be indicating in the green arc. An operating cylinder head temperature gauge is not required for VFR flight, but if the plane has one, its indication should be noted at this time. Also check any other engine gauges that happened to be installed in the airplane (fuel flow, fuel pressure, EGT).

There is one exception to the oil-pressure limitation mentioned above. If you fly an airplane powered by a normally aspirated Lycoming engine, and the outside air temperature is below about 45 degrees Fahrenheit, and the engine's oil is stiff (perhaps because you're not using multigrade oil and/or didn't obtain pre-heat), oil pressure may be near redline during the runup. This is normal for this type of engine, during cool weather. As mentioned in Chapter 3, the oil-pressure sensing probe for Lycoming engines is placed quite near the oil pump, and indicates relatively high to begin with. On cold days, with cold oil in the sump, indicated pressure will often be close to redline until the oil warms up and thins out. In a non-turbocharged Lycoming, it is acceptable to initiate the takeoff roll with the oil pressure at, or even a needle width over, redline, *as long as the oil temperature needle has come off the peg.* If the oil temp gauge still hasn't come to life, wait a few minutes before taking off.

Notice that the above comments do not apply to turbocharged engines. Turbocharger bearings receive (engine) oil through relatively tiny feed holes and must have both significant *pressure* and *oil flow* for proper lubrication to occur. Even more important, turbocharger controllers and hydraulic wastegate actuators (in planes that have automatic boost controls) also depend on proper oil flow for normal operation. The metering orifices in automatic controllers are quite tiny, with the result that cold, thick (congealed) oil has a hard time flowing to the wastegate in quantities sufficient to allow proper wastegate functioning on takeoff. In other words, if you attempt to take off with cold oil, you may overboost the engine, as a sluggish controller (filled with cold oil) tries in vain to keep up with throttle movements in the cockpit. In a turbocharged aircraft, *always wait for the oil-temperature needle to come off the peg before initiating takeoff.* Adequate oil *pressure*, alone, is not sufficient to ensure proper boost control.

If it seems, on cold mornings, as though it takes an unusually long time for oil temperature to come up, don't be surprised. The oil temperature gauge, remember, does not begin at zero; the "cold" end of the scale (usually not numbered) starts at something like 60 degrees Fahrenheit. The needle won't come off the peg until oil at the probe reached 65 or 70 degrees F. On a cold morning, that could take a while.

Every year, at least two or three pilots make the mistake of hurrying to the end of the runway, performing a quick runup, and taking off with oil so cold that the engine oil pump cavitates on rollout, leaving the surprised aviator with a seized engine somewhere over the departure end of the runway. Don't be one of these pilots.

Mag Drops

You may have noticed, if you've spent any time comparing owner's manuals, that different operating handbooks often specify different acceptable rpm drops on single-mag operation. Where one handbook specifies a maximum single-mag rpm drop of 125 rpm, another may allow 175 rpm. Some manuals, such as that for the 1968 Cessna Skylane, specify no particular rpm drop at all, but allow no more than 50 rpm difference *between* mags. What's particularly curious about this state of affairs is that often, different manuals (with different advice) refer to airplanes with identical magnetos. Who then—the inquisitive pilot might ask—are we to believe?

The fact of the matter is, the maximum allowable mag drop is arbitrary. In practice, the actual magnitude of the rpm loss on single-mag operation depends on many variables, including oil temperature, mag-to-engine timing, internal mag timing, runup rpm, density altitude, and even whether the aircraft is faced upwind or downwind for the check. What's important, therefore, is not that you adhere mindlessly to someone's cookbook formula for doing a mag check, but that you understand the purpose of the check and the factors that influence its outcome, so that you can draw your own conclusion as to the airworthiness of the ignition systems. Unfortunately, many pilots have lost sight of the basics in this department.

The main purpose of the mag check is to ascertain whether each magneto, independent of the other, is capable of sustaining ignition. It's actually a check of the *total* ignition system, of course, including not just the magnetos, but the spark plugs, ignition harness, and ignition switch. Obviously, if any of these items were to prove defective, it would show up in the mag check. Still, you might ask: Why not try the mags out individually at idle rpm? Why "run up" at all? Surely there is something to be said for the fact that the magnetos must be considered functional if the engine is running well enough to allow you to taxi to the end of the runway. There's only one flaw with this line of reasoning. The voltage output capabilities of (and demands on) a magneto increase exponentially with rpm. Your mags' performance at idle rpm says nothing about how they'll perform at or near takeoff rpm. Similarly, the optimal *ignition timing* for your engine is different at low rpm than at high rpm; your engine will start easily and idle smoothly, for example, with mag timing dangerously retarded. You would pick this up as a too-large mag drop during a runup at 1,800 rpm. (On takeoff, the engine might even backfire and run poorly.) You'd miss it entirely if you did your "mag check" at 1,000 rpm.

The actual choice of rpm for doing the mag check, as said before, is not terribly important; anything between 1,600 and prop redline will do. The important thing is to be consistent from runup to runup, so that trends are easily monitored. (It's no use trying to compare a 150-rpm mag drop that happened yesterday at 2,000 rpm with a 100-rpm drop that occurred this morning at 1,700 rpm.) Also keep in mind that single-mag rpm drops will tend to be larger on cold mornings than on warm-weather days, and drops tend to be smaller when facing into a stiff wind than when facing downwind. (Obviously, anything that tends to increase internal friction or prop loading will accentuate rpm drop on single-mag runup.)

When conducting the actual check, the correct procedure is to switch quickly from "both" to "left" (or "right"—it doesn't matter), leave the switch on the individual mag for about five seconds *or until rpm stabilizes*, then switch quickly back to "both" before trying the other magneto in identical fashion. Avoid leaving the switch on a single mag for periods longer than 15 seconds, if you can, since the inactive spark plugs tend to load up quickly with fuel and/or oil during the brief shutdown of one mag. Also avoid switching directly from "left" mag to "right" mag without going through "both." You want to let the engine rpm stabilize on "both" before testing *either* mag; this gives you a valid basis of comparison of the two mags.

(Also, you want to let the just-shut-down plugs burn off for a second or two before trying them on single-mag operation.) *Terminate the mag check immediately if loud backfiring or extremely rough operation occurs on any magneto.* If these conditions exist, you aren't going flying (if you're in your right mind)—so relax and taxi back. Don't encourage exhaust-system, accessory, or engine mount damage by continuing to run on a dangerously rough mag.

Get in the habit of performing the mag check with your ears, rather than your eyes. Don't fixate on the tachometer. (Don't ignore it, either; use it to set the initial runup rpm, and take note of the actual rpm drop on each mag.) Your ears will, in most cases, do the best troubleshooting and make the final go/no-go decision. If your tach ever fails, you should still be able to do a competent runup, by ear. (Manifold pressure indications are virtually worthless during a no-tach runup.) What you want to hear, of course, is a smooth, very slight, rpm drop on single-mag operation, with no popping or sputtering, or other unusual noises.

Generally speaking, an rpm drop of up to 200 rpm on either mag should be of no concern if the drop is smooth, close for both mags, and consistent with past runups. If the plane just came out of maintenance, on the other hand, and both mags now show a 200-rpm drop where yesterday it was 75 rpm, then something may be wrong. Use common sense. The actual amount of rpm drop is not as important as smoothness, consistency between mags, and consistency from one runup to the next. Trend monitoring is the key. If your mags have always given a 75-rpm "split" between them, and still do, fine. But if your mags used to have a 10- or 20-rpm difference, and in the last few flights you've watched the spread go to 100 rpm, it's time to call in a mechanic.

Consistency between mags is important because each magneto is a kind of laboratory "control" for the other; both are subject to the same environmental effects, the same stresses, the same wear rates, the same rpms, etc. So if one mag suddenly begins to deviate from the other on runup, you can be fairly sure something is amiss (so to speak) either with the magneto or its associated harness and plugs.

Consistency between mags is also important from an ignition-timing standpoint. Remember that either magneto, alone, can initiate combustion. (Each magneto fires a complete complement of spark plugs—one in each cylinder of the engine.) Thus, if your left mag, say, is firing the spark plugs too soon—a condition known as early, or "advanced," timing—then combustion begins at an inappropriately early point in piston travel, *regardless of what the other magneto is doing.*

Combustion begins when the earlier of the two magnetos "fires," period. This is a very significant point to ponder, for several reasons. First, advanced timing is extremely destructive to engines (high-compression engines particularly); the high cylinder head temperatures and combustion pressures generated can liberate heads from barrels, pistons from rods, etc., and destroy valves and other components. Secondly, resetting mag-to-engine timing in the advanced direction is—unfortunately—a fairly common bootleg maintenance procedure used to correct large rpm drops on runup. (Continental reported finding some years ago that a high percentage of engines received at the factory for rebuilding had timing set as much as 20 degrees in advance of the properly advanced setting.) Honest mechanics also sometimes make honest errors in setting timing—timing marks on crankshaft flanges, etc., are often hard to read correctly—and some "early drift" also occurs through normal wear of breaker points inside the magneto. But perhaps the most insidious thing of all about advanced timing is that most pilots ignore the warning signs. *The initial indication of improperly advanced timing is a smaller-than-normal rpm drop on runup.* Far from being something to brag about, a mag drop of less than 50 rpm is genuine cause for concern—it means mag-to-engine timing needs to be looked at immediately.

Retarded timing (or "late" timing) is generally less destructive than advanced timing, but still spells trouble since it robs an engine of power output. (Most aircraft engines are timed at around 22 degrees BTC—that is, plug firing is set to occur at 22 degrees of crankshaft rotation before top dead center of piston travel on the compression upstroke. Best power actually occurs around 30 degrees BTC for most aircraft engines, but engine makers prefer to trade efficiency for cooling—cylinder heads run cooler and last longer with timing at 20 or 22 degrees—and anyway, some safety margin must be allowed for normal point drift, careless maintenance, bad gasoline, and/or other eventualities. To find the proper timing specs for your engine, look on the metallic engine data plate riveted to the top of the crankcase.) Most pilots are wary of retarded timing, since it results in a large rpm drop on runup.

By now it should be evident that if there is no roughness on either mag (indicating possible plug or harness problems), the single-mag rpm drop can tell you valuable things about ignition timing. It can also tell you whether your P-lead wiring is intact. The magneto primary lead (P-lead) is the wire that runs from the condensor section of the mag to the cockpit switch. When this wire is grounded through the switch, the magneto is rendered inoperative. Conversely, when the P-lead is not connected to ground, the mag is "hot." If you should note a *zero-rpm* drop on single-mag operation, it means you have a broken P-lead *and the mag is hot at all times, even with the switch "off."* P-leads break with fair frequency (due to normal vibration), so it pays to be alert to this possibility.

If the rpm drop on single-mag operation is accompanied by roughness in the engine, chances are good you've got an intermittently firing spark plug. (Whether the spark plug itself is intermittent, or the harness wire associated with it is bad, is another question.) Engine are normally set up to run very rich at idle—rich enough, often, to foul plugs while taxiing. If you suspect a fouled plug is the cause of a large mag drop, repeat the mag check at a higher rpm (such as 2,100), and if it still gives a large drop, lean the mixture about a third of the way for ten seconds or so (with the switch on "both") before repeating the check once more. If this brief burn-off doesn't clear the plug(s), and your mag drop is still unacceptably high (200 rpm or more), you're probably better off taxiing back to the ramp than ground-running the engine for an extended period in an attempt to burn off the errant igniters. (Extended ground running creates localized hot spots under the cowling, encourages prop erosion, and—in general—doesn't do anything good for the airplane or engine.)

A multi-probe exhaust gas temperature system can be a great troubleshooting aid in a situation like the one just described. If you suspect poor combustion in one or more cylinders, all you have to do is take a look at EGT indications for each individual cylinder. (Exhaust gases are hot enough to provide useful EGT readings at runup rpm in most cases.) Any cylinder with aberrant EGT indications can be considered to harbor one or more bad plugs or ignition leads. The cross-check provided by your other "good" cylinders is, of course, what makes the multi-probe EGT a fabulous troubleshooting device in a situation like this.

What if you don't have a multi-probe EGT? A single-probe EGT will give you *some* information for troubleshooting (you may be able to isolate the problem to one bank of cylinders, for example, or—if the probe is located on one riser only, rather than at a cluster point—you may be able to definitively rule out one cylinder as the problem zone). For sake of argument, though, let's assume you have *no* EGT system at all. And you have a rough mag with a 200-rpm drop. What then?

The thing to do is turn the ignition switch back to the rough (bad-sounding) magneto, and taxi back to the ramp. Allow the engine to run for about two minutes on the bad mag. (It will shake, but unless there is bad backfiring, let it idle roughly for at least a minute.) Shut the engine down with the mixture control, then get out and open the cowling. Before the engine cools down, quickly test each exhaust riser for warmth by touching a plastic pen (or other suitable object) to each pipe and noting whether the plastic tends to melt. All but one of the exhaust pipes will be hot enough to melt the plastic. The cold riser leads to the jug with the aberrant plug (or ignition wire).

You can perform this so-called "cold cylinder test" by laying hands (or thermometers) atop the cylinder cooling fins, if you want. Old-time mechanics simply spit on each exhaust riser, figuring that the one that doesn't hiss is connected to the cold jug.

In any event, when you've spotted the cold cylinder, trace each ignition wire from that cylinder back to the respective magnetos. Needless to say, the wire that's connected to the "bad mag" is also connected (at the other end) to the non-firing spark plug—whether it's top or bottom—in the cylinder in question. Now you'll be able to tell your mechanic "We've got a bad bottom plug in cylinder number three," or whatever, and you can be on your way in no time.

It occasionally happens, of course, that neither spark plugs nor magnetos are to blame for an ignition problem. Harnesses do break down. When a harness breakdown occurs, it's usually at the elbow or bend in the wire just outboard of the spark plug terminal end, where vibration (and abuse suffered in maintenance) is maximal. Another spot where ignition-wire trouble occurs is at the magneto end of the wire, where contact is sometimes lost between the contact pin and the spring in the distributor output section. Trouble here can often be spotted visually after the harness is uncoupled from the distributor (the back plate comes off quickly by undoing four screws); a charred-looking or pitted contact pin is the tipoff.

Then too, it sometimes (not often, however) happens that dew or rain collects inside a magneto to produce a rough engine during runup. The mag in question may fire very poorly until the engine comes up to normal temperatures (after prolonged idling), after which the roughness mysteriously disappears (because the water has been baked off). The symptoms are very similar to those produced by a stuck valve, in that the engine starts out rough but eventually smooths out before takeoff. Of course, with a stuck valve, operation is rough on *both* mags, until the valve loosens up. My inclination would be to shut down any engine that does not run smoothly within 30 seconds of startup. Any question of whether wet mags or stuck valves are to blame can be resolved by shutting the engine down and (very cautiously) turning the prop over by hand, feeling for compression. If a valve is hung open, one jug will feel totally flat.

CHAPTER SIX

TAKEOFF
AND CLIMB

Takeoff and Climb

Takeoff and climb are the most critical regimes of aircraft engine operation. Robust engine performance, needless to say, is essential for a safe departure. Mismanagement of power controls (or misinterpretation of trouble signals) during the takeoff roll can spell disaster, for a single-engine airplane *or* a twin.

In very simple airplanes (such as trainers with carbureted engines), takeoff operation consists of little more than pushing the throttle in and leaving it there until the plane reaches cruise altitude. For more complex aircraft, such as a Pressurized Centurion or A36TC Bonanza, there are many more things to consider, such as fuel flow adjustments (via boost pump and/or manual leaning), precise manifold pressure settings, and engine instrument indications (including, perhaps, TIT or multi-channel EGT/CHT). Even in the simplest airplane, however, there is a right way and a wrong way—or at least an intelligent way and a stupid way—of conducting the takeoff and climb with regard to engine operation.

First things first. Always comply with the pre-takeoff checklist requirements given in the aircraft operating manual. (No attempt will be made here to duplicate this information, since it varies so much from plane to plane.) At the very least, this means going through the runup procedure described in the previous chapter, turning on the boost pump as applicable, opening cowl flaps (if present), and checking all power gauges for function. In cold weather it is particularly important to check the oil temperature indication. If the oil-temp needle hasn't come off the peg, it's advisable to delay the takeoff, especially if you're using straight-weight (non-multigrade) oil, and especially if your engine is high-compression, high-horsepower, and/or turbocharged. Why take a chance on oil-pump cavitation? Pull over to the side of the runway and wait until the oil temp needle begins to rise—then take off.

Regardless of what else is on your (handbook) checklist, *listen* to the engine as you pull onto the runway, and *feel* the throttle response as you taxi. Try to establish a non-visual sense of whether the engine *feels* right for takeoff. (This will free up your eyes and hands to do the things they're best at, such as set the DG and clock, turn the transponder on, and ensure that the prop, mixture, and carb heat controls are full-forward.) Mentally prepare yourself for an aborted takeoff. (Twin-engine pilots are already in this habit; it doesn't hurt for single-engine pilots to do this, too.)

When you're in position and cleared to begin the roll, open the throttle *smoothly* to about the one-third-power position; listen for smooth acceleration. As you open the throttle beyond the one-third point, you should be off the brakes—not so much because you want a cooling airflow to begin through the cowling (which you do, of course, but the cooling effect—which, remember, increases with the *square* of airspeed—is virtually nil until airborne), but rather to minimize prop bending loads and abrasion damage. When you run the engine up while standing on the brakes, the prop tips—which produce most of the thrust—actually *bend forward* as much as an inch or two (depending on how large your prop is). If you've ever wondered how NTSB investigators can tell, from a badly mangled prop, whether an engine was putting out power when it hit the ground, this is how: Prop tips curled *forward* indicate that thrust was being produced; tips curled *back* indicate that the prop was turning slowly enough when it hit for the blades to be bent backwards on impact.

Apply the last two-thirds of throttle smoothly and slowly, but not pitifully slowly; once the engine is out of the idle range, acceleration should be good. As you come in with the throttle, *check the engine instruments.* Oil pressure, in particular, should be seen to hold steady (or increase to a constant value, less than redline) as power is applied. If any unusual noises are heard, or the oil pressure hiccups, abandon the takeoff at once, even if it means locking the brakes, ruining the tires, and going off into the overrun area.

In a complex airplane, certain additional duties may be called for: In a Mooney 231 or Piper Turbo Arrow, for instance, manifold pressure must be set manually by the pilot on takeoff, using approximately half of throttle travel (it varies from day to day). In these airplanes, the fixed-wastegate turbocharger system allows the engine to develop more than redline manifold pressure on takeoff, and the pilot must modulate power output manually. This is also somewhat true for such airplanes as Cessna's Turbo Centurions and the Beech Turbo Bonanzas, which—although they have automatic controllers (in contrast to the Mooney 231)—tend to exhibit some fluctuation in setpoint MP on each takeoff. On the first takeoff of the day, particularly in cold weather, manifold pressure may stabilize at a value higher than redline; and the pilot must be alert to this fact.

Leaning for Takeoff

As mentioned before, in some airplanes—such as the A36TC Bonanza—the fuel flow must be set manually during the takeoff roll, further compounding the pilot's duties. There are times, too, when pilots of smaller airplanes will need to adjust fuel flow on takeoff—namely, on hot days or at high-elevation fields. Fuel-metering systems are, as a rule, set quite rich in the full-throttle mode (frequently as much as 50 percent beyond best-power mixture), thanks to the *economizer* or *power-enrichment* feature of the carburetor or fuel injector. The last inch or so of throttle travel activates the fuel-enrichment valve or jet, sending a virtual fire-hose blast of extra fuel to the engine. The result, ironically, is that manual leaning of an aircraft engine at full throttle actually results in more than full rated power being developed—more than 100-percent power. It also results in overheating (and quite possibly, detonation). The excess of fuel is there for a reason: namely, cylinder cooling.

On a hot day, or on takeoff from a field with an elevation several thousand feet above sea level, your full-throttle mixture is not just overrich, but super-overrich (because the incoming air is "thinner," or less dense in oxygen), further degrading the engine's already poor power output at altitude. Under such conditions, it is entirely permissible—indeed, highly advisable—to lean on takeoff. In fixed-pitch-prop airplanes, the standard technique is to taxi into position, set the parking brake, run up to full power, and lean until the rpm peaks. This actually produces a leaner fuel-air ratio than is normally gotten during full-throttle operation at sea level, however, so for cooling purposes it is probably better to lean to peak-rpm (which is actually best-power; see Chapter 8), *then enrichen again* until the rpm stabilizes at a point midway between the no-leaning and best-rpm leaning points. (If obstacle clearance or density-altitude climb performance is critical, of course, you can sacrifice engine cooling and lean to max-rpm.)

In normally aspirated high-performance airplanes with constant-speed props, takeoff leaning is definitely also advisable under high-and-hot conditions, but here it's not possible to lean with reference to rpm, since prop governing keeps rpm constant. The ideal thing to do is lean with reference to the EGT gauge; briefly pull the mixture back until EGT peaks, then enrichen until the EGT is 125 to 150 degrees (roughly) on the rich side of peak. (This is best-power mixture.) If a multi-channel EGT system is present, lean by reference to the leanest cylinder (not the hottest cylinder; there's a difference—see chapter eight). Detonation should not be much of a problem if your mag timing is set correctly, since the engine is not developing sea-level power, and you are only leaning to "best power" mixture, not peak EGT. (Some detonation margin exists between best-power and peak.) Nonetheless, do not dally while finding peak EGT.

If you have neither an EGT nor a fixed-pitch propeller, lean (on hot/high days) by reference to manifold pressure, or—if one is available—the fuel-flow gauge. Power available decreases three percent for every 1,000-ft increase in density altitude. Therefore, it's a simple matter to calculate what percentage of your normal sea-level fuel-flow is appropriate for a given density altitude. Let's say you are taking off from El Paso on a 6,000-foot density-altitude day, and your normal sea-level fuel flow on takeoff is 178 pounds per hour. The horsepower degradation due to density altitude in this example will be about 18 percent. You want your fuel flow to be about 18 percent lower than normal; hence 82 percent of 178 pph, or 146 pph, is the desired fuel-flow.

If your aircraft is turbocharged, you are getting more or less full power on takeoff under almost any pressure-altitude conditions; hence there is no real need to lean on takeoff, ever (unless otherwise specified in your operating handbook, of course). This is not to say that most turbocharger systems compensate for *density* altitude, however; most, in fact, do not. They merely provide a certain amount of manifold pressure. (Pressure, rather than density, is the key word here.) Some handbooks, accordingly, advise the pilot to manually adjust fuel-flow and/or MP up or down to specific values on hot/high takeoffs (an example is the Cessna P210N manual). Two airplanes on which bonafide density-compensating turbo systems *can* be found are the Piper Turbo Aztec and Navajo Chieftain.

Reduced-Power Takeoff

Pilots frequently ask whether it isn't advisable, for the sake of engine longevity to take off with something less than full throttle as a routine operating procedure. After all, it is often remarked, maximum engine wear occurs at full power; by avoiding full power except in emergencies, it ought to be possible to cut engine wear by a significant margin—and thereby extend TBO. Many high-performance engines come with a placarded 5-minute takeoff-power limitation. Why not take off at max-continuous power (instead of the 5-minute rated T.O. power), and be done with it?

If your airplane is blessed with what you think is a surplus of unneeded power—and you live in an uncongested area, with clear, unobstructed terrain at the departure end of either runway—I'm not going to tell you that you can't safely take off with less than 100-percent power. People take off with less than 100-percent power every day in places like Denver and Albuquerque. Obviously it can be done.

I don't recommend it, however. First of all, despite all the talk of maximum wear occurring at full power, I have never heard of any engine making significant TBO gains solely on the basis of reduced-power operation; nor have I seen evidence that airplanes which make greater-than-normal numbers of full-power takeoffs (trainers, for instance) get *less* than normal TBO. If anything, just the opposite seems to be true. People who run their engines hard seem, as a group, to have fewer problems reaching TBO than people who "baby" their engines—at least at the Cessna 150/172/182 level. (Obviously, crop dusters, airshow pilots, and pylon racers fall into a different category altogether.) Owners of 470-series Continentals, in particular, seem to encounter more problems with oil consumption, cylinder varnish, ring sticking, etc. at low power than when engines are run "hard." (Once rings stick, of course, rapid barrel wear occurs, especially at the top of the cylinder.) The argument that low-power operation, in general, can't *hurt* an engine is demonstrably false.

Consider what happens to engine cooling when a reduced-power takeoff is conducted. The last inch or so of throttle travel, remember, engages the power-enrichment feature of the carburetor or fuel injector, sending extra fuel to the cylinders for cooling. This is extra fuel that (generally speaking) you can't get any other way. At full-rich mixture, a Lycoming O-320-E (with MA-4-SPA carburetor) has a full-throttle specific fuel consumption of 0.57 lbs/bhp/hr, at full 2,700 rpm. Reduce power to 2,500 rpm, however (or approximately 75 percent), and the specific fuel consumption goes down to 0.53 lbs/bhp/hr—again with full-rich mixture. The difference of seven percent is due to power enrichment. (Volumetric efficiency also enters in here, but for the O-320-E, volumetric efficiency is only about one-percent better at 2,500 rpm than at 2,700 rpm, so the vast majority of the change in s.f.c. is due to mixture enrichment.) While it may sound small, a seven-percent change in fuel flow can have a noticeable effect on CHT. Of course, the problem is complicated by the fact that your cooling airflow is—in all likelihood—also less, since you are not developing full power for the rollout and initial climb. The bottom line is that engine cooling *is affected* by the decision to use reduced power for takeoff. It's entirely possible to see *higher* cylinder head temperatures (and *more* engine wear) on a partial-power takeoff than on a normal, full-power takeoff.

But even if you could save fuel, extend TBO, *and* properly cool your engine on takeoff by using partial power, there's another good reason not to do it. Namely, safety. Any way you look at it, your performance is going to be less on a reduced-power takeoff than it would be on a full-power departure. You'll break ground further down the runway; your altitude will be less as the end of the runway goes by; your initial rate of climb will be diminished; and you'll gather airspeed less quickly. One minute after brake release, you won't be anywhere close to where you would normally be. And yet, statistically, the first minute after brake release represents one of the most significant "risk windows" in all of flying. If an engine problem is going to manifest itself, chances are good it will do so in this minute—and probably after you have broken ground. Where do you want to be when this happens? Personally, I want to be as high as possible, with as much airspeed reserve as possible, so that I'll have as much time as possible to collect as much of my wits as possible before landing again.

There are only two occasions when you should consider using less than full power for takeoff. One is if the density altitude is below sea level. Obviously, if density altitude is abnormally *low*, you can expect the engine to develop more power than it otherwise would—again, approximately three percent more per thousand feet. To keep from operating the engine at greater than rated power, you may have to ease off the throttle slightly. (The effect will not be great, however, unless you are in a truly cold part of the planet.)

Another occasion calling for reduced power is any time you have reason to believe the gasoline is not of sufficient octane rating. Perhaps you are visiting a foreign country or are in a remote region where the only available fuel is of poor quality. As mentioned before, detonation (knocking) is not normally something you can hear in an airplane; you have to go by EGT, CHT, and/or "feel." If any of these suggest combustion knock on takeoff, by all means throttle back, just as you would do to stop knocking in a car. (Be sure, also, that the prop is in the full high-rpm position, the mixture is rich, carburetor heat is off, and both mags are working.)

Climbout

Climb operation in a trainer is simple: Just hold the throttle all the way in and trim for Vy. In virtually all other aircraft, however, climb brings with it additional power-management considerations. If the engine has a takeoff-power rating as well as a maximum-continuous-power rating, a power reduction must be made after takeoff. (Such a

Initial power reduction should not be made until the aircraft is "cleaned up," and at a comfortably high altitude (noise abatement procedures notwithstanding).

power reduction is also usually made with other engines, although it is not required.) Prop rpm may need to be adjusted as well as manifold pressure; and fuel flow may need to be reset with the mixture control. The boost pump, if present, will at some point need to be turned off, and in many aircraft cowl flaps require attention. In addition, turbocharged airplanes with manual wastegates will require wastegate adjustments in the climb. Multiply all this by two, for a twin.

Pilots of airplanes with constant-speed props (or engines with more than 180 horsepower) have long been taught to make a power reduction almost immediately after takeoff—ostensibly to save wear and tear on engines, although also admittedly to save wear and tear on the pilot's ears—even though, as stated above, most engines do not carry a time limitation on takeoff power ratings. For most airplanes, maximum continuous power is also "rated power" (takeoff power), and there is no real need to reduce power during climb. Of course, in any normally-aspirated aircraft, a power reduction of three percent per thousand feet occurs *automatically* anyway, due to thinning of the atmosphere. So a power reduction—in the strict sense—is inevitable.

Opinions vary on when to begin the power reduction. Some pilots begin it while the airplane is still in ground effect above the runway; others wait until leaving the airport traffic area (or the county or state). Noise abatement procedures notwithstanding, the author's firm advice is to wait until a safe altitude has been reached to make the power reduction—if indeed you choose to make one at all (other than that due to atmospheric pressure lapse). There is nothing unsafe about continuing to run an aircraft engine at redline power. To the contrary, statistics indicate that engine mechanical problems most often manifest themselves during *the initial power reduction*. This in itself is reason enough to delay power reduction until a safe return to the airport (or to a suitable forced-landing site) can be made. In some cases—such as when you are breaking in a freshly overhauled (or topped) engine—you may choose not to make a power reduction at all.

Pilots of so-called high-performance aircraft are often squeamish about leaving the "balls to the wall" for more than a minute or two after takeoff, when actually there is no legitimate reason to be squeamish at all (unless of course CHT or oil temperature are approaching redline). In certification testing, engines are required to spend many hours at full power, with CHT and oil temperatures a specified amount *above redline*. (FAA also requires a complete teardown inspection after 150 hours of test-stand operation; and generally speaking, during the teardown very little wear is usually discovered.) Helicopter piston engines—which are identical to fixed-wing engines except for strengthened valve springs, modified cam-lobe profile, and minor changes to the lubrication system—spend essentially *all* of their time at full throttle, with crankshafts typically turning 200 rpm over what you and I would consider a decent redline. Copter pilots think nothing of it. Why should you worry about leaving your throttle in for a few minutes after takeoff?

When a safe altitude has been reached (generally traffic-pattern altitude or higher), you can begin the power reduction by retarding the *throttle* first, then the prop pitch. Then, if necessary, you can adjust the mixture control and turn off the boost pump. The only exception to the sequence is if you are flying an older airplane powered by a geared Lycoming engine (a Twin Bonanza, say), where sudden unloading of the crankshaft relative to the large, gyroscope-like propeller can cause detuning of crankshaft counterweights. In Aero Commanders, Queen Airs, Twin Bonanzas, and other airplanes powered by GO-, GSO-, or IGSO-series Lycoming engines, the standard operating practice is to reduce *prop rpm* first, then throttle back—which, of course, is exactly counter to tradition. (This happens to be why Beech put prop levers where throttles normally are in its light twins, in case you were wondering.)

If you are unsure about how much of a power reduction to make after takeoff, consult your operating manual. Generally, the drill is to throttle to 75 percent, or to the top of the green arc(s) on the manifold pressure and/or rpm gauges, as applicable. (Naturally, you what to avoid continuous operation in any yellow arcs.) In a normally aspirated plane, a typical initial power setting is 25 inches of MP and 2,500 to 2,700 rpm.

During the climb, of course, you will lose about one inch of manifold pressure for every 1,000 feet of altitude gained (except in certain turbocharged aircraft), which means you will need to push the throttle back in a little now and then as you continue to climb.

Fuel Flow Adjustment

The question of whether and how to lean during climb is an important one for many operators of high-performance aircraft, whose engines may run smoothly over a very wide range of fuel flows, some of which are undoubtedly harmful to the engine over the long run. Some operating handbooks (particularly for fuel-injected engines) call for leaning to specific fuel-flow values during the climb. Obviously, if this is the case, one need only follow the guidelines spelled out in the factory manual. (With experience, those guidelines can be modified to suit the operator's needs.) In other instances, the proper procedure is anything but obvious.

It's interesting to note that some manuals (e.g., Cessna P210N) advise the pilot never to lean at more than 80-percent power, while others (Cessna 177B) say not to lean during climb below 3,000 feet; while still others (Grumman AA1A) say to begin leaning at 5,000 feet. (Then too, many manuals are completely silent on the subject.) To some extent, this merely reflects the conservative bent of the

In the Beech Queen Air (and other planes with geared Lycoming engines), the standard procedure is to reduce rpm before reducing throttle, not the other way around.

manufacturers, who are of course interested in keeping service claims (and lawsuits) down. But it also reflects real differences between engines, and specific installations. A Pressurized Centurion's turbocharged TSIO-520-P engine, with its redline manifold pressure of 38 inches and its solid-stemmed exhaust valves, is not going to tolerate aggressive leaning as well as a Cardinal's normally aspirated O-360 Lycoming, which struggles to get 28 inches of manifold pressure and has cool-running sodium-filled valves. Different considerations apply to each engine.

The problem of how to lean during climb is made much more manageable by a multi-probe EGT system, especially if used in conjunction with a fuel-flow totalizer/computer. At low altitude, the chief limitation to leaning can be considered to be detonation onset or the attainment of damaging exhaust-system temperatures, whichever occurs first. (At high altitude, the principal limitation is apt to be CHT redline.) Detonation onset is difficult to measure directly and cannot be predicted precisely, owing to differences in cylinder head design, fuel blending, mag timing, etc. Damagingly high EGTs are easier to measure, but only if one's EGT or TIT gauge is calibrated in actual degrees Fahrenheit (or Celsius); 1,600 degrees F can be considered a benchmark temperature, above which rapid heat damage to exhaust valves, turbochargers, risers, clamps, mufflers, is virtually assured.

No one piece of advice is going to suffice for every engine type, under every set of environmental circumstances, but some general rules hold:

—If your engine is so rich that it's beginning to run rough, you can and should lean it during the climb. This is true for just about any engine type.

—If you have a multi-probe EGT system, lean to best power mixture on the leanest cylinder (see Chapter 8), after making your initial power reduction. Best power is approximately 150 degrees rich of peak.

—If you have a single-probe EGT, lean to approximately 200 degrees rich of peak. Do not spend any more time than is necessary finding peak, particularly with a turbocharged engine. (Write down fuel flow values for future reference.)

—The above advice aside, *do not exceed 1,600 degrees EGT, nor redline CHT, under any circumstances.*

These guidelines apply at any altitude, for virtually any engine. Do use common sense, however. A turbocharged, geared, fuel-injected engine is going to be less forgiving of abuse than a small, low-output, carburetor-equipped engine. Also, an engine with solid-stemmed exhaust valves and bronze guides (i.e., early Continental IO-470) is going to be less heat-resistant than one with sodium-filled valves and Ni-Resist guides (late Lycoming O-540).

Cowl Flaps

The use of cowl flaps bears on the question of how to lean during climb (above), inasmuch as cooling airflow through the cowling can—within reason—be substituted for cooling fuel flow through the cylinder intake ports. The important thing to remember about cowl flaps is that their effectiveness is related not only to how much you open or close them, but to indicated airspeed. (Indicated airspeed and cooling are both functions of ram air pressure.) Another worthwhile thing to bear in mind about cowl flaps is that they affect not only CHT, but oil temperature, and the temperature(s) of everything under the cowling, as well.

Keeping the above facts in mind, it is nonetheless true that cowl flaps are your *primary* means of controlling CHT. The CHT is a slow-to-react gauge, but the effect of cowl flaps is immediate: open them, and you send a torrent of cool air through the cylinder cooling fins. (The air may not seem cool if it's a summer day and you're at low altitude; but remember, the wind chill factor at 100 knots is considerable!) What, you might ask, is a good target temperature (a good ''CHT-ref'')? Lycoming redlines its cylinder heads at 500 degrees F, while Continental sets 460 F as redline. Nevertheless, as Lycoming has pointed out on many occasions (in its service literature), cylinders will generally last longer if maintained at no more than 435 degrees. This is sound advice, and it applies just as surely to Continental cylinders as to Lycoming's. Cylinder-head metallurgy isn't that different from one engine to the next.

One may well ask whether it is better, from a fuel-economy standpoint, to open the cowl flaps and lean in climb, or close the cowl flaps (thus permitting a faster ROC and shorter time-to-climb) while enrichening the mixture to keep CHT in the green. In other words, is it more efficient to throw gasoline at an engine (for cooling), or throw air at it? We assume here that CHT is near redline (or capable of getting there quickly), since otherwise—if you have plenty of ''green arc'' left—it doesn't matter anyway. The answer to this question is not at all obvious when you consider, for example, that cowl flaps are remarkably ineffective on some airplanes, but quite effective on others. (Also, cooling drag is a large fraction of total drag on some airplanes; less so on others.) There are additional subtleties involved, as well. Leaning to best power in the climb may tend to raise CHT, but it also provides you with about five percent more power (''best'' power), which can of course be converted into airspeed for cooling. The answer is probably best arrived at by test-flying. If opening your cowl flaps has only a small effect on your rate of climb (but noticeable effect on CHT), you are probably well advised to lean the engine during climb, keeping cowl flaps open to limit CHT. If on the other hand you find that opening the cowl flaps cuts your rate of climb in half (not likely, except perhaps in a normally aspirated airplane at high altitude), you will probably find better efficiency in keeping cowl flaps *closed* and mixture *rich*.

In any event, remember always (regardless of CHT indications) to avoid EGTs in excess of 1,600 Fahrenheit when leaning during climb. EGT and CHT), are not always well correlated with each other (as we'll see in Chapter 8).

Cruise-Climb

It is common practise, in high-performance aircraft especially, to choose an airspeed well in excess of Vy for climb—not just to improve engine cooling, but for better visibility straight ahead and lessened time enroute. Vibration and cabin noise may be reduced as well.

The effect of increased airspeed on engine cooling is pronounced. The reason is that cooling effectiveness is related to cooling drag, and cooling drag—like any other drag—increases with the square of (indicated) airspeed. Thus if you double your airspeed, you will quadruple your cooling drag (in theory, at least).

Bear in mind, then, that you have at least three ways of dealing with a too-high CHT: mixture control (enrichen), cowl flaps (open), and airspeed (increase). You also can reduce the throttle—or open it, as the case may be. If your throttle is within an inch of being all the way in, try pushing it in all the way (with mixture full forward) to activate the power-enrichment feature. If all else fails, of course, simply reduce power.

Boost Pump Usage

Some auxiliary fuel pumps are capable of affecting fuel flow when turned on or off, while others are not. Therefore it is important to monitor the fuel flow gauge (or EGT, if no fuel-flow instrument is installed) to determine whether mixture should be readjusted after turning the boost pump off in climb.

When should you turn the boost pump off? Follow the airframe manufacturer's advice (as given in the aircraft owner's manual). But also take into consideration other factors. If you are going on up to oxygen altitudes, leave the pump(s) on throughout the climb. Temperatures generated ahead of the firewall during the climb may cause heat soakback into fuel fittings, flow dividers, etc. resulting in possible vapor formation at high altitude. (Remember that the lower atmospheric pressure at high altitude increases the tendency of liquids to boil.) On reaching cruise altitude, turn the boost pump(s) off only after engine compartment temperatures have stabilized for five minutes or more.

Automotive gasolines often have a higher vapor pressure than aviation gasoline, so if you are using car gas (under an STC, presumably), plan on using the boost pump a good deal of the time. Vapor formation is most likely to occur on the ground with a hot engine during long holds for takeoff, and during extended climbs to altitude. High OAT (outside air temperature) is an aggravating factor.

Needless to say, the boost pump should be activated any time fuel pressure fluctuates for any reason in flight. In addition, the pump should be on whenever switching tanks, except in such aircraft as the Cessna 310, which have boost pumps located inside the main tanks but not in the auxiliary tanks. In the 310 (and other tip-tanked Cessna twins), switching from tips to aux tanks cuts out boost pump action. If fuel-pressure fluctuations are noted, the pilot must switch back to the tips before boost pump action can be had.

CHAPTER SEVEN
CRUISE
OPERATION

Cruise Operation

Aircraft engines are generally operated between 55 and 75 percent power in cruise—not just because "that's the way it's always been done," but for some very good practical reasons. The fact is that above 75 percent, fuel consumption in the "full rich" mixture position is very high, yet the mixture cannot be leaned without producing detonation or cylinder overheating. (There's also the simple fact that, for a normally aspirated engine, it's impossible to pull more than 75 percent power at the most commonly used cruise altitudes.) Below 55 percent, on the other hand, airspeed falls off so rapidly that the airplane's usefulness as a transportation tool is thwarted.

Maximum aerodynamic efficiency, in terms of minimum power required to sustain flight (and therefore best endurance), occurs somewhere around 30 or 40 percent power, depending on the airframe/engine combination. This is not the "least drag" condition, however; going slightly faster will actually reduce drag and give best L/D (lift to drag ratio) speed, which is also the maximum-range condition. This occurs at around 35 or 45 percent power (sometimes higher; again, it depends on the airframe-powerplant combination). Most pilots are interested in maximizing the ratio of fuel economy (miles per gallon) to time enroute, however, which is a very different thing from maximizing "miles per gallon" (range) by itself. That, in a nutshell, is why most cruising today is done at not much less than, nor a great deal more than, 55 percent power. Fifty-five percent provides the best tradeoff between miles-per-gallon and time enroute. You can go faster, of course, by selecting 65 or 75 percent; but fuel consumption goes up much quicker than airspeed. (Power-required increases as the *cube* of airspeed, in fact.) Conversely, if you throttle back from 55 to 45 percent, you will very likely increase mpg economy by 10 percent but increase time-getting-there by something like 20 percent. The actual numbers vary from plane to plane, and there are many subtle complications involved (e.g., variations in volumetric efficiency with throttle setting; variations in low-speed aerodynamic efficiency from one airplane design to the next; altitude effects; etc.), but the net effect is nearly always the same: You get the best transportation value from an airplane by cruising at or near 55 percent.

Some of these concepts are perhaps best illustrated by Fig. 7-1, which shows miles per gallon (fuel efficiency) plotted against airspeed (which is of course a function of power) for a Mooney 201 operated at five, ten, and fifteen thousand feet. The graph also shows CAFE scores plotted against airspeed for the 201. The Competition for Aircraft Flying Efficiency, an efficiency race held every year in California, uses an efficiency formula which, simply stated, multiplies speed, payload, and miles per gallon together. The idea is for contestants to try to maximize fuel efficiency and speed (and payload) at one and the same time. In Fig. 7-1, you can see that the Mooney 201's maximum-fuel-efficiency airspeed—i.e., where the "mpg" curve peaks—is 120 miles per hour, at 5,000 feet. This is, of course, the best-range mark, and it corresponds to about 35 percent power. (The airspeed is somewhat greater at higher altitudes, simply because 35 percent power produces higher true airspeeds in the thinner air.)

The best CAFE efficiency—the best *trade-off* of airspeed and fuel efficiency—occurs, for the Mooney 201, at about 145 mph TAS at 5,000 feet; 160 mph at 10,000 feet; and 180 mph at 15,000 feet. The corresponding power settings range from about 45 percent to 55 percent. (At 15,000 feet, 55 percent power is, of course, all that's available to the normally aspirated Mooney 201.) These curves will vary somewhat from one plane to the next, and with gross weight. But the bottom line, again, is that the best way to optimize fuel efficiency and time efficiency *simultaneously* is to operate at or near 55 percent.

Except for the fact that some engines can't tolerate leaning at 80 or 90 percent power (while others foul spark plugs and accumulate cylinder varnish at very *low* power settings), there is no fundamental engineering or maintenance reason why you can't select any power setting you want between 40 and 100 percent for cruise. It's essentially an arbitrary decision, based on operational considerations peculiar to the mission at hand. If you're a commercial operator flying an electric billboard at night, and you want to maximize time spent aloft, you'll cruise at a lower power setting than you would if you wanted to get from Omaha to San Antonio in the least time possible. Factor in headwinds, and everything changes further. (Best mpg-economy per time-enroute will occur at something greater than 55 percent going into the wind, and at less than 55 percent going downwind.)

The choice of cruise power, then, is (and should be) based on practical considerations, rather than anything said in this book. Different operators have different cruise-power needs. The purpose of this chapter is to advise the reader, *after* cruise power needs have been determined, of the possible TBO and maintenance ramifications

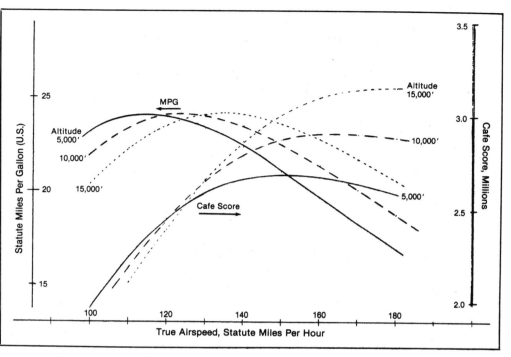

Figure 7-1: Mpg efficiency (and CAFE efficiency) versus true airspeed for various altitudes, in the Mooney 201.

of various cruise-configuration decisions. In addition, the basic "how to" of setting an engine up for cruise will be discussed, for normally aspirated as well as turbocharged powerplants.

Leveling Off

Most instructors now teach their students to level off by lowering the nose of the airplane first, then reducing power as necessary to establish the desired cruise-power setting. The order of events is certainly correct as described, but the manner in which the level-off is carried out should be a function of prevailing engine-instrument indications. If oil and/or cylinder head temperatures have risen to within 10 percent of redline, for example, it is advisable to allow airspeed to build to a high value before reducing power; also, a cooldown period of a minute or two should be allowed before closing any cowl flaps. This is particularly true at altitudes above 13,000 or 14,000 feet, where indicated airspeeds are low and engine cooling is slow to occur. In a turbocharged aircraft, a typical level-off sequence might be to trim nose-down as necessary to remain at one altitude; wait for airspeed to rise to the *anticipated final cruise indication*; bring

115

the throttle back to the desired manifold pressure; reduce prop rpm as necessary; allow IAS, CHT, and oil temp to stabilize for a few minutes, then close the cowl flaps in stages. (In many turbocharged airplanes flying at high altitude, manifold pressure will be affected by changes in airspeed, rpm, etc., necessitating a second and sometimes a third or fourth run-through of the above sequence, after the gauges have stabilized. See discussion further below.)

As always, the proper sequence of power-knob operation is to *reduce manifold pressure first, then bring prop rpm back.* Conversely, when increasing power, adjust prop rpm first, then bring in the manifold pressure. (Use care when moving the prop control in the "increase rpm" direction, to prevent surging.)

If the airplane has two engines, it's best to synchronize the propellers by establishing a "master" and a "slave" engine. When the master engine is set at the desired rpm, sync the other engine to it by *reducing* rpm. (In other words, when setting up for cruise, initially put the master engine at a slightly lower rpm than the slave engine.) Prop governing is much better dampened in the "reduce rpm" direction, and props are easier to sync if the slave engine's rpm is being adjusted *downward.* (If the airplane is equipped with an electronic propeller synchrophaser, one can turn the unit on after manually syncing the props in the manner just described. Turn the prop sync *off,* however, prior to any large power changes.)

Prop pitch and manifold pressure are, of course, interactive, so it may be necessary to go through more than one iteration of the basic cruise setup sequence even in a non-turbocharged aircraft. As you reduce rpm in a C/S-prop aircraft, manifold pressure will tend to rise, often as much as an inch for every 200 to 300 rpm of prop-speed reduction. Conversely, an increase in propeller rpm will have the effect of diminishing manifold pressure slightly. With practice, of course, one can anticipate and compensate for these effects, initially setting cruise MP slightly lower than MP-ref, for example, so that when rpm is brought back, manifold pressure increases to the desired final value.

Interpreting Power Charts

Finding out exactly what combination of manifold pressure and rpm will produce a given power setting for cruise, at a give pressure altitude and OAT, is not always easy. The airframe manufacturers' owner's manuals give this information in the form of cruise-performance charts or "power setting tables." The tables are usually presented for just a few altitudes (multiples of 2,500 feet, for example, up to service ceiling) and a few temperature conditions, and generally only one mixture setting. Fuel flows may or may not be given. The drawback to these charts is that, because they *are* charts (not graphs), quite a bit of interpolation skill is called for in non-standard conditions, or at non-standard altitudes. Also, the owner's manual charts are not very good at answering such questions as: "What's the most efficient combination of rpm and MP for my engine?" And: "To what degree am I safe operating the engine in an 'over-square' condition?"

What it comes down to is this: The owner's-manual charts will get you in the right ballpark for cruise, but for precisely determining power settings, you need detailed altitude-performance graphs such as those given in the engine manufacturers' operator's manuals. (If you don't already have an operating manual for your engine, see your local Lycoming or Continental dealer. Lycoming's operating handbooks are $5.00 each as of this writing, while Continental's are $10 each.)

Fig. 7-2 shows sea-level and altitude power curves for the Continental O-470-R, a typical carburetor-equipped general-aviation engine. (The power graphs for other engine models—including fuel-injected ones—take the same general shape.) Fig. 7-2 actually consists of two graphs that can be analyzed separately. The left-hand portion is a simple plot of horsepower against manifold pressure at sea level. The various sloping lines represent the power available at various propeller rpms as MP is increased. The lines end at the right where full throttle has been applied.

One fact that should be immediately obvious is that the engine produces its highest horsepower at the highest prop rpm (2,600 in this case). Power, by definition, is the rate at which work is done; and obviously an engine does its work fastest at high rpm. This is not to say that prop efficiency, volumetric efficiency, or fuel efficiency are at their peak at redline rpm (quite the contrary)—merely that if what you need is raw power, the prop control should be all the way forward. For takeoff and initial climb, of course, that's where the prop knob is.

Something else you should note from the left-hand portion of Fig. 7-2 is that while power generally increases as manifold pressure goes up (for any given rpm), maximum horsepower does *not* occur at the maximum attainable full-throttle manifold pressure. The max attainable manifold pressure of 29.2 inches occurs at the lowest rpm setting—1,800 rpm. At redline rpm (2,600), the full-throttle manifold pressure is barely 28.0 inches—almost an inch and a half less. The reason for this is that the engine's volumetric efficiency decreases as rpm increases (with the throttle full-open). Volumetric efficiency is

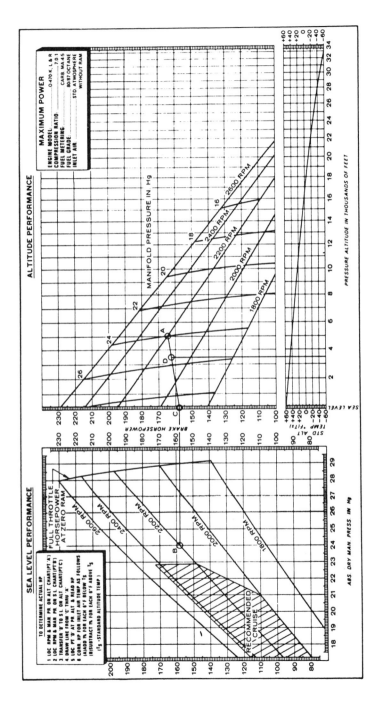

Figure 7-2: Sea-level and altitude performance charts for the Continental O-470-K/L/R series. Note that maximum sea-level horsepower occurs at maximum rpm (even though more manifold pressure is available at lower rpm). The manufacturer has clearly indicated the limits of "oversquare" operation by the crosshatched area on the left. Temperature correction: Add one percent power for each 6 degrees F below standard temperature; subtract one percent power for each 6 deg. above standard.

the ratio of the actual volume of air pumped by an engine on one piston stroke, to piston displacement; it's essentially a measure of the air-breathing efficiency of an engine. Since piston engines burn mostly air (15 or 16 parts of air to one part fuel), volumetric efficiency is critical to a reciprocating engine's performance. (Volumetric efficiency can be artificially increased—by supercharging—to well over 100 percent, but for a normally aspirated engine a V.E. of 75 or 80 percent is typical.) The decrease in volumetric efficiency with increasing rpm is simply a reflection of the fact that at higher rpm, the velocity of the air into the engine increases, with attendant gains in air friction, air-filter "delta-p" (pressure drop), and gas-inertia effects at the intake ports. To a great extent, these factors apply to all other piston aircraft engines; however, valve timing and induction system design vary significantly from model to model, and it is interesting to note that the full-throttle MP "spread" (relative to rpm) for many small Lycoming engines is only half an inch or so, whereas for most Continentals it is more than an inch.

In Fig. 7-2, the manufacturer has chosen to state unambiguously exactly where the bounds of "recommended cruise" power settings are. For the O-470-R, recommended max cruise comes at approximately 22.75 inches of manifold pressure and anywhere from 2,200 to 2,450 rpm, with downward adjustments to MP called for as rpms less than 2,200 are selected. Does this mean that other combinations of MP and rpm *can't* be used? No. Not unless the cockpit gauges have red or yellow arcs in the non-preferred zones (which they don't, in the Cessna 182). The manufacturer's recommended cruise settings, in this case at least, are just that: recommendations.

So much for sea-level power output. To determine power output at altitude, a second graph is needed—in this instance, the right-hand portion of Fig. 7-2. The O-470 "altitude performance" graph contains lines of constant rpm (sloping off to the right), plotted against pressure altitude, again with horsepower on the vertical scale. Lines of constant manifold pressure are drawn more-or-less vertically. They represent, in effect, the maximum attainable manifold pressure for the given altitudes, and not surprisingly (since manifold pressure is limited by ambient atmospheric pressure), they get smaller and smaller as the graph is read further and further to the right, until at 16,000 feet you're lucky to get 16 inches of manifold pressure. Notice that the manifold pressure lines do not go exactly straight up and down, but shift slightly as they cross the various rpm lines. This is in keeping with our observation earlier that volumetric efficiency is better at lower rpms. Indeed, you can see that for any given altitude, a higher manifold pressure is available at a low rpm than at a high rpm. (More *raw power* is available at high rpm, however.)

The right-hand portion of Fig. 7-2 essentially allows you to determine exactly what percentage of power you're developing at any altitude (at full rich mixture). All you have to do is locate your manifold pressure and rpm on the left-hand (sea level) graph, draw a horizontal line to the "horsepower" (vertical) scale, locate your MP and rpm on the right-hand graph, and draw a line connecting that point to the horsepower number you just found; then read straight up from the proper altitude (along the bottom of the right-hand graph) until you encounter the angled line you just drew. That will give you your actual horsepower (according to the vertical axis) at altitude—for standard-day OAT. If outside air temp is not standard, add one percent to actual power for every 10 degrees Fahrenheit (6 Celsius) below the standard-day OAT indicated on the temperature scale underneath the altitude-performance graph, or subtract one percent power for every 10 degrees above normal. (This is an approximation. The actual correction factor by which to multiply standard-day horsepower is the square root of the standard-day OAT divided by the square root of the actual OAT; but both temperatures must be in degrees Rankine—i.e., Fahrenheit degrees above absolute zero. Add 460 to Fahrenheit to get degrees Rankine.) The answer you get, in horsepower, can of course be converted to percent power by dividing by your engine's rated power. Why the manufacturers don't simply plot the vertical scale(s) in percentage power is unknown.

For most engines, a given combination of MP and rpm results in slightly more horsepower at altitude than at sea level, for reasons having to do with charge density and exhaust back-pressure (prop efficiency can enter in, too). The effect is seldom large, however—usually around five percent for normally aspirated engines under typical conditions—and can actually be nonexistent for some turbocharged engines.

Notice that the altitude performance charts do not correct for ram air pressure (if any) against the intake scoop. Depending on airframe design, this can be a substantial effect (as in the Cessna 310) or a negligible one (Piper Seneca). Some idea of the magnitude of ram effect can be obtained by switching to alternate air—if your plane is so equipped—in cruise flight, while watching manifold pressure. Usually the drop on alternate air is less than an inch. For low-horsepower, low-speed aircraft, it can be considered so slight as to be negligible.

Note that higher OATs tend to reduce power, while lower OATs increase it, to the extent of one percent per 6-degree (F) change.

Choice of RPM

Pilots often ask whether it is best (for the engine; for fuel efficiency) to cruise at a high rpm and a low manifold pressure, or a low rpm and a high manifold pressure. (Both can result in the same power output.) The arguments typically offered for low rpm/high MP are that it reduces friction and wear (extending TBO) while favoring fuel efficiency (due partially to better prop efficiency); also, noise is usually reduced. The arguments *against* low rpm/high MP usually center around "lugging" the engine, adverse vibrations (some engines do not run smoothly at rpms below 2,400), possible problems with detonation on leaning, and the age-old "oversquare" taboo. Pilots for many years have been taught a rule of thumb which says that in order to avoid damaging the engine, one should avoid manifold pressures which are greater than the crankshaft rpm in hundreds. That is to say, a "square" condition—e.g., 24-square (24 inches and 2,400 rpm)—is acceptable, whereas an "oversquare" condition, such as 24 inches and 2,100 rpm, should be avoided. Exactly when and where this myth got its start is anybody's guess. The fact of the matter is that almost every turbocharged aircraft in the world operates "oversquare" 90 percent of the time, and no one would dare suggest that any supercharged engine be operated differently. Indeed, most normally aspirated engines are flown "oversquare" on takeoff (if not also cruise), routinely. No one is suggesting, surely, that any other kind of takeoff is safer or better?

There is nothing magical about operating in a "square" fashion—in fact, there is no intrinsic relation between rpm and manifold pressure that favors any particular correlation of numbers. If prop speed were measured in revolutions per second (or per fortnight), there would be no "squaring" of crank speeds with manifold pressures—just as there could be no squaring-up of MP and rpm if the former were measured in kilopascals or psi. Your propeller has no idea what the pressure is inside your intake manifold. (Colonel Lindbergh demonstrated this to good effect in World War II with the P-38. By demonstrating that the P-38's engines would not destroy themselves on long-range flights at ultra-low rpms, Lindbergh spawned a change in standard operating procedures that saved U.S. forces many thousands of gallons of gasoline.)

Ultra-low-rpm operation sometimes—but not always—results in better fuel specifics (see below). But it should be pointed out that while friction is somewhat less at lower rpms, combustion pressures are higher (all else being equal) and piston ring wear can thus be considered roughly the same at low-rpm cruise as at high-rpm cruise (at a given percentage of power). Likewise, valve temperatures go up at low rpm (again, at a constant power), accelerating erosion and mitigating any supposed TBO-extending effect. Also, some engines vibrate alarmingly at low rpms, almost certainly to the detriment of TBO. In short, the maintenance benefits of low-rpm operation are largely, if not totally, illusory.

The limiting factor in using low rpm in conjunction with high manifold pressure is apt to be cylinder overheating and/or detonation, from an operational standpoint; from an engineering standpoint, the limits are best described in terms of bmep (brake mean effective pressure). Bmep can be thought of as the average pressure developed inside the combustion chamber. It is directly proportional to horsepower output (bhp) and inversely proportional to rpm. (It is also inversely proportional to displacement—which is just another way of saying that at a given level of horsepower production, bmep is higher in a low-displacement engine than in a large engine, all other things being equal.) Generally speaking, as bmep goes up, the octane requirement of the engine increases. In automotive parlance: the more you lug the engine, the likelier it is you'll begin to knock.

If you want to calculate bmep for a given engine, or a given operating condition, you can do so quite easily with your pocket calculator. Bmep, in psi (pounds per square inch), is equal to brake horsepower times 792,000, divided by the product of rpm and displacement.

$$\text{bmep} = \frac{792{,}000 \times \text{bhp}}{\text{rpm} \times \text{cu.in.}}$$

Aircraft engines with takeoff bmeps of under 140 psi (which includes, for example, the Lycoming O-320-E series and low-compression O-540 models) are capable of running on 80/87 avgas. Engines certified for 91/96 avgas generally have bmeps in the 140 to 150 psi range; the requirement for 100-octane gasoline comes at around 160 psi bmep. These "break points" are only approximate; there are exceptions to them. (Cylinder head design has a profound influence on knock tendency. Angle-valve cylinders, for instance, tend to run knock-free to higher bmeps than parallel-valve heads, even at the same displacement, same compression ratio, etc.) But the point is, bmep is what you're affecting when you alter the relationship between MP and rpm at a given power setting. You're less likely to knock—and therefore can lean more aggressively—with a high-rpm power setting than with a high-MP, low-rpm setting.

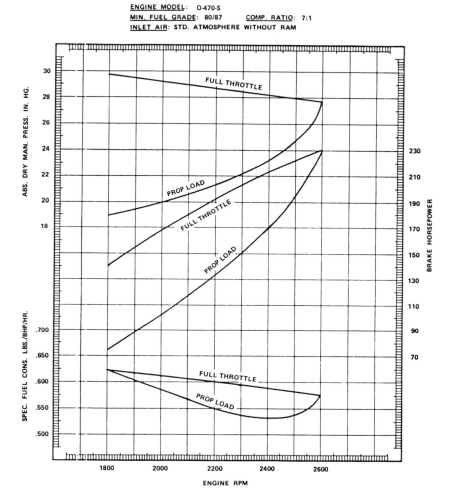

ENGINE MODEL: O-470-S
MIN. FUEL GRADE: 80/87   COMP. RATIO: 7:1
INLET AIR: STD. ATMOSPHERE WITHOUT RAM

Figure 7-3: Sea-level manifold pressure, horsepower, and specific fuel consumption versus rpm (O-470-S).

All of this is not to suggest that operators should set rpm and MP randomly, or in contradiction to manufacturers' recommendations (or in contradiction to common sense). It's worth emphasizing, however, that (contrary to hangar mythology) if there are no yellow or red arcs on the tach or MP gauge in the operating region under consideration—and if the airplane is not placarded to the contrary—it is perfectly acceptable to operate continuously at any combination of rpms and MP that will get the job done.

The optimum-sfc rpm varies, naturally, from engine family to engine family. (It also varies with throttle position and altitude.) Specific fuel consumption is affected by intake tuning and camshaft design (volumetric efficiency), compression ratio, and many other factors. Practical considerations—such as exhaust-temperature limitations—often keep one from operating at theoretically attainable specific fuel consumptions. Nonetheless, it is helpful to know the best-sfc rpm(s) for a given engine; and this information is usually available in the engine manufacturer's operating handbook. (It is *not* usually contained in the airframe manufacturers' operating manuals. Owner's-manual power charts frequently insinuate that at a given power setting—65 percent, say—fuel flow will be the same at *any* combination of MP and rpm, which is seldom, if ever, true.)

Unfortunately, the engine-maker's handbooks don't usually plot prop-load sfc for various cruise altitudes, so it's impossible to know for sure the ideal rpm and manifold pressure combination to use at altitude. (With an accurate fuel-flow computer on board, it's a fairly straightforward—if tedious—task to determine optimum settings by flight testing. All you'll need are the sea-level and altitude power curves for your engine, so that you can set power accurately at altitude, and a note pad to record rpm, MP, altitude, TAS, and fuel flow. Later, you can make plots of miles per gallon versus rpm for various power settings—55, 65, and 75 percent, for example.)

Some engine manufacturers' handbooks contain part-throttle fuel consumption graphs (Fig. 7-4) which show lines of constant-power and constant-rpm plotted against fuel consumption on the vertical scale, and manifold pressure along the bottom. On such graphs, look for the spot in the constant-power curve where the curve bottoms or achieves a minimum with regard to fuel consumption. Reading from that part of the curve, you'll get both a manifold pressure and an rpm for ''best sfc.'' Generally, the rpm is quite low.

The question is: What rpm/MP combination *will* ''get the job done'' most effectively? How does one determine the combination that gives the best fuel efficiency? The longest TBO?

The fuel-efficiency question is not terribly hard to answer. Most engine operating handbooks come with sea-level performance curves of the type shown in Fig. 7-3 (which is actually three graphs in one). At the top of this graph, we see how manifold pressure varies (along the vertical axis) with engine rpm (bottom axis). One line—labelled ''Full Throttle''—shows the relationship between MP and rpm as throttle setting is held constant, while the other curve—''Prop Load''—shows how MP varies with rpm as the throttle setting is

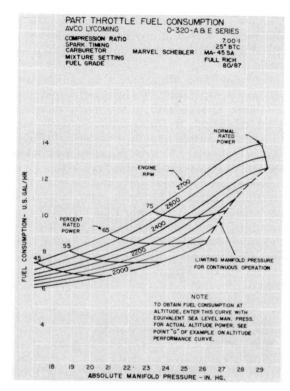

PART THROTTLE FUEL CONSUMPTION
AVCO LYCOMING   O-320-A & E SERIES

COMPRESSION RATIO		7.00:1
SPARK TIMING		25° BTC
CARBURETOR	MARVEL SCHEBLER	MA-45 SA
MIXTURE SETTING		FULL RICH
FUEL GRADE		80/87

NORMAL RATED POWER

ENGINE RPM
2700
2600
2400
2200
2000

FUEL CONSUMPTION - U.S. GAL/HR

PERCENT RATED POWER
75
65
55
45

LIMITING MANIFOLD PRESSURE FOR CONTINUOUS OPERATION

NOTE
TO OBTAIN FUEL CONSUMPTION AT ALTITUDE, ENTER THIS CURVE WITH EQUIVALENT SEA LEVEL MAN. PRESS. FOR ACTUAL ALTITUDE POWER. SEE POINT "G" OF EXAMPLE ON ALTITUDE PERFORMANCE CURVE.

ABSOLUTE MANIFOLD PRESSURE - IN. HG.

Figure 7-4: Part-throttle fuel consumption for Lycoming O-320-A and -E. Lines of constant power connect the various rpm curves. The effect of various choices of rpm and manifold pressure on fuel consumption can be determined quickly with the aid of this kind of graph.

varied while the prop control is held full-in. The middle part of the graph plots horsepower against rpm in the same constant-throttle-versus-varied-throttle fashion. The bottom set of curves are of the most interest to us; they tell us how specific fuel consumption varies with rpm (again for throttle held full-in, or for throttle being varied with prop pitch at highest-rpm), *at sea level*. Specific fuel consumption is a measure of how efficient an engine is in converting gasoline (or other fuel) to usable power. It is expressed, quite logically, in pounds of fuel per horsepower per hour (or pounds per hour, per horsepower, if you prefer). With the throttle wide-open, the O-470 has a sfc of about 0.60 lbs/hp/hr, rising slightly as rpm is reduced and falling slightly as rpm is increased. The curve we're interested in (since we're not operating at full throttle, normally, but partial throttle) is the bottom one, shaped like a bathtub. This curve of sfc versus rpm with varying throttle shows a definite minimum at 2,400 rpm for the O-470-R & S (at which point the engine is developing a sea-level manifold pressure of 23.2 inches, according to the topmost prop-load curve of Fig. 7-3). What it means is that at any rpm much above or below 2,400, the fuel efficiency of the engine will be less than optimum, at least at sea level.

Part-throttle power curves are also often useful for determining where the limits of "oversquare" operation lie. Again, though, such limitations—when present—are always also given in the form of placards or instrument-dial indications in the cockpit.

As a practical matter (without getting heavily into flight testing or power-curve analysis), it's usually best, from a fuel-efficiency standpoint, to simply choose the lowest cruise rpm consistent with manufacturers' MP/rpm limits, on the one hand, and engine vibration, on the other. The author's Turbo 310 (Continental TSIO-520-B engines) operates smoothest at 27 inches MP and 2,200 to 2,250 rpm, which happens to be around 55 percent power—and that's where cruising is done. On the other hand, the author has flown some airplanes (with 470-series Continentals) whose engines were not smooth-running at rpms below 2,400. I recall a G-33 Bonanza, in particular, that always seemed happiest at 2,450 rpm—no less. That, consequently, is where I cruised it.

Remember one thing: Don't place *too much* emphasis on power charts and handbook numbers (not even this handbook!)—because common sense dictates that an engine (any engine) is going to experience the least stress, and give the longest TBO, at that rpm that gives the least vibration (all other things being equal). How much vibration is too much? This is admittedly a subjective matter. Go ahead and rely on your best judgment, but also bear in mind that long, thin, flexible crankshafts (such as found on 360-series Continental sixes, and Lycoming IO-720 eight-cylinder models) are generally less tolerant of vibratory abuse than short, stiff crankshafts (as on four-cylinder O-200 Continentals and O-360 Lycomings). In twin-engined airplanes with four-cylinder Lycomings, engine mount cracking (caused by poorly dampened engine vibration) can be a significant concern. Modify your cruise rpms accordingly.

Generally speaking, in normally aspirated airplanes, the best procedure—once an efficient rpm has been selected on the basis of power charts or flight tests—is to fly at an altitude where wide-open throttle gives the desired percentage of power (55 percent, 65, or whatever). Why wide-open throttle? Because a wide-open butterfly valve presents the least restriction to incoming air and—all other things being equal—nets a significant gain in volumetric efficiency.

In turbocharged airplanes with manual wastegates, the same advice holds true. Climb until the throttle is full-open before closing the wastegate(s); then select a cruise rpm in the best-sfc range. If the turbocharger wastegate is controlled by an automatic controller (rather than manual verniers), full-throttle operation may not be practicable. In either case, however—manual wastegate or automatic—it is generally true that the very best fuel specifics come when the wastegate has closed all the way. In an automatic-controller-equipped plane, this will be below critical altitude if less than full throttle is applied. (Partial-throttle critical altitude and full-throttle critical altitude are two different things. The latter occurs at a higher altitude above sea level.) How do you know if you have reached the partial-throttle critical altitude (i.e., wastegate fully closed) in an automatic-controller-equipped plane? Momentarily retard the prop rpm and watch what happens to manifold. If manifold pressure falls off immediately, your wastegate is already closed all the way. If it holds steady as you begin retarding prop rpm, the wategate hasn't yet closed 100 percent.

Since a turbocharger uses, in effect, "free" waste heat energy (from the exhaust) to dramatically improve the engine's volumetric efficiency and power output, one might be tempted to think that a turbocharged engine would operate most efficiently with the wastegate always closed, no matter what the altitude or throttle setting. This is not the case, however. A closed wastegate represents a significant restriction in the exhaust—and the increase in back-pressure is *not* compensated by the increase in "front pressure" from the compressor as long as the throttle is closed. If the throttle is closed, even partway (and it will be, if you're below critical altitude), the compressor's output is simply being thwarted by the throttle butterfly. So you're actually making the engine work *harder* by closing the wastegate before opening the throttle. If you have the option, *always open the throttle before beginning to close the wastegate.* Otherwise you'll needlessly increase the thermal and mechanical loadings on your engine.

Turbocharged Engines at High Altitude: Special Considerations

When the wastegate is open, a turbocharged engine behaves much the same way as a normally aspirated engine (not surprisingly, since the turbo engine is essentially normally aspirated under this condition). But when the wastegate is closed, the T-engine behaves differently, in many respects, than the unblown engine. For the most part, we're talking here about flight at oxygen altitudes (over 14,000 feet), although in many turbo installations, wastegate closure can occur even below 10,000 feet. In an automatic-controller system, it is very often the case that when the pilot throttles back from climb power to cruise power manifold pressure, levelling off at an altitude below the airplane's stated critical altitude, the turbo wastegate(s)—having been partway open during the climb, to prevent overboosting—will close all the way as the throttle is reduced for cruise. The altitude at which the wastegate fully closes is thus dependent on throttle setting. At full throttle, the point of wastegate closure defines the airplane's "critical altitude." Part-throttle critical altitude is a different matter. (If you're ever in doubt as to whether your wastegate has closed all the way, momentarily reduce prop rpm and observe what happens to manifold pressure. If manifold pressure tends to remain constant, the wastegate was not closed before; if instead it drops off with rpm, the wastegate was and is closed.)

When the wastegate is closed and all exhaust must pass through the turbocharger on its way out the engine, a closed feedback loop is formed, such that any increase in fuel or air flow through the engine tends to increase the turbocharger's output, further increasing manifold pressure and flow, adding to turbine speed, etc. Conversely, a decrease in mass flow through the engine causes the turbine to slow down, with the result that turbocharger output falls off, further reducing flow through the engine. At low altitudes, these "bootstrap effects" are not as apparent, because (with an open wastegate) much of the exhaust flow is diverted around the turbocharger, and in an automatic-controller system the controller itself (usually nothing more than an aneroid bellows sensing upper deck pressure, and controlling oil flow to the wastegate by means of a poppet valve) functions like a household thermostat to maintain the manifold pressure where the pilot selects it. Above the critical altitude, the controller is effectively taken out of the loop. (It will open the wastegate if an overboost condition develops, but otherwise its "thermostatic" feedback function is eliminated.) Hence considerable potential for manifold pressure fluctuations exists once the turbocharged aircraft has reached high-altitude cruise.

That's not all. In a turbocharged aircraft, the fuel pump output (and the injector system's output) is regulated in part by turbocharger output pressure (upper deck pressure). This is necessary, of course, to keep the turbocharged engine from leaning out as the turbo compressor comes on line. But the net effect is that the fuel pump is part of the turbo feedback loop, too—which means that the potential exists not only for wild manifold pressure fluctuations, but for large fuel-flow excursions at altitude, as well.

What does all this mean in practice? First of all, it means that the MP/rpm interaction the exists for normally aspirated planes (that is, the tendency for manifold pressure to increase as rpm is decreased, and MP to decrease as rpm is increased) will also exist for turbocharged planes *at low altitudes*—but when the turbocharged plane has arrived at a high enough altitude to bring about wastegate closure, the MP/rpm effect will reverse polarity. In other words, when the wastegate is closed, a decrease in rpm will result in a *decrease* (not an increase) in manifold pressure, while an increase in rpm will cause an *increase* in MP—just the opposite of what normally occurs down low.

The fact that fuel flow is part of the turbocharger feedback loop means that changes in fuel flow caused by leaning (or even caused by turning boost pumps on or off) can have a noticeable effect on manifold pressure. When the mixture is leaned, for instance, fuel flow decreases significantly (see next chapter). This in turn reduces the mass flow through the turbocharger, reducing turbine speed and diminishing compressor output, with a corresponding loss in manifold pressure. The magnitude of the MP change varies from installation to installation; however, it is often large enough to require resetting the throttle.

Ram effects are also important. Changes in airspeed can affect air pressure at the engine air-inlet scoop, causing manifold-pressure excursions downstream. Of course, ram air pressure at the air filter or air scoop is not usually very high; but at critical altitude, the outlet/inlet pressure ratio developed by the turbo compressor may be as high as 3-to-1 (meaning that the turbocharger acts to multiply small pressure changes). If the turbo is operating at a pressure ratio of 2.5-to-1, say, then a one-inch ram pressure change at the intake air scoop is converted to a two-and-a-half-inch change in pressure inside the upper deck. (If the throttle is open all the way, manifold pressure will increase 2.5 inches, accordingly.) A very small increase in airspeed can thus easily translate into a one- or two-inch manifold-pressure change. This effect is very noticeable during the initial level-off period, as airspeed builds (and also during the initial phase of letdown). Obviously, it must be compensated for by the pilot.

The net result of all this, as you can well imagine, is that setting a turbocharged engine up for high-altitude cruise can be a time-consuming process, involving as many as five or six iterations of the basic throttling-back process to get MP, rpm, and fuel flow to end up exactly where the pilot wants them. Moreover, once set up for cruise, the pilot must monitor airspeed, fuel flow, throttle position, and rpm for any changes which might set off excursions in manifold pressure. (Once a change is noted, a *small* correction in manifold pressure can be made with the throttle. It is important, however, to wait a minute or two between changes, to let the gauges settle down to their new values, before making additional corrections. Otherwise, you can easily spend the whole flight "chasing the gauges.")

Large spontaneous manifold-pressure excursions are possible in some aircraft, under high-altitude cruise conditions. Such bootstrapping episodes can be alarming (and frustrating), since they come and go without warning and—more often than not—can't be counteracted easily by the pilot. The secret is usually to increase rpm. The worst bootstrapping generally happens at rpms below 2,400 (in direct-drive engines); so the minute severe bootstrapping is encountered, a higher rpm should be selected. (Keep an eye on manifold pressure, however, and maintain it within the limits for the aircraft, which are often posted in placard form on the panel.) In some cases, service kits are available to relieve bootstrapping by means of larger-diameter wastegates, improved controllers, etc. (See your dealer or write to the airframe manufacturer for details.) Besides increasing the crankshaft rpm of the engine, or operating at a lower altitude to open the wastegate, not much can be done about bootstrapping, unfortunately. For turbo operators, it comes with the territory.

CHAPTER EIGHT
MIXTURE
MANAGEMENT

Mixture Management

Combustion is the heart and soul of transportation, indeed of twentieth-century civilization itself. All practical forms of locomotion depend on combustion. The combustion of wood drove the first steam engines; coal propelled the railroads and built nations. Alcohol and kerosene (it is now widely forgotten) powered the first internal combustion engines built by Otto and Langen. Gasoline, originally a waste product of kerosene production which was customarily dumped into waterways or burned in the open air, had not long been a popular fuel when the Wright Brothers first flew with it. It is now, of course, more popular.

Combustion is basic to the operation of all heat engines; it is the *sine qua non* of turbojets, Wankels, diesels, and two- and four-stroke cycle engines alike. When it comes to getting the most out of an internal combustion engine, understanding the combustion process is half the battle. Since the mixture control affects combustion directly (more directly, in fact, than any other cockpit control), an understanding of combustion is central to effective mixture management.

If you remember nothing else from this chapter, remember this: Combustion is defined as the spontaneous rearrangement of a fuel (gasoline, for instance) and oxygen (e.g., from the atmosphere) to give carbon dioxide, water, and energy. The energy liberated in combustion, of course, is not "created" out of nothing; it was there to begin with in the chemical bonds in the fuel and oxygen. The rearrangement of the fuel and oxygen atoms to form compounds with lower-energy chemical bonds (H_2O and CO_2) is what allows energy to mysteriously appear in the form of heat, light, and gas expansion. The more "oxidizable" the fuel, generally speaking, the more heat energy you get at the end of combustion. For instance, a hydrocarbon such as methane, CH_4, can rearrange with four oxygen atoms to give one CO_2 and two water molecules—liberating about 12,500 calories of energy per gram of methane in the process. Carbon monoxide (CO, sometimes known as "producer gas") will also burn, but since it is already in a partially oxidized state, not much energy will be released. In fact, one CO molecule can react only with one oxygen atom to give one CO_2 molecule—liberating 4,900 calories per gram of fuel. This is the key to understanding why, for example, alcohols make for relatively poor fuels on an "energy per pound" (calories per gram; BTUs per pound) basis. Methanol's chemical formula is CH_3OH. (All alcohols, by definition, contain an "OH.") The presence of an oxygen in the fuel molecule means less energy will be released on combustion.

One thing you should bear in mind is that regardless of the fuel, combustion—when carried to completeness—always yields carbon dioxide and water (in the form of steam) as the end products. Even an engine that belches smoke and soot and runs so rich that gasoline can be smelled in the exhaust, is producing mostly water and CO_2 (believe it or not). The same is true of living organisms, which in a crude sense are nothing but heat engines. Fuel (food—carbohydrates mostly) is recombined in the body with oxygen to give CO_2 and water, and energy. Biochemists refer to the combustion process in living organisms as "respiration." The end products, in any case, are the same. (Interestingly, the efficiency of biological systems in converting fuel energy to usable energy for growth and locomotion is only about 35 to 40 percent—that is to say, 60 percent or more of the energy released in digestion and respiration is liberated as waste heat. This is not far different from the best efficiencies attained by heat engines.)

Gasoline is a complex mixture of hydrocarbons, very difficult to characterize in a few sentences other than to say that the molecules are of a high-energy-density configuration (mostly long-chain, linear molecules with an average carbon-chain length of eight), with few or no naturally occurring alcohols, and a minority, at best, of unsaturated cyclic C_6 compounds (generically known as "aromatics"). What's important to remember is that the makeup of gasoline is such that a relatively large amount of oxygen must be reacted with it to give sustainable combustion. In fact, the ideal chemical ratio of oxygen to gasoline—what chemists refer to as the "stoichiometric" ratio—is 3.37 mass-equivalents (grams or pounds) of oxygen for every one mass-equivalent (gram or pound) of fuel. But air is not 100-percent oxygen; in fact, at sea level, it is only 22 percent oxygen. So the chemically correct ratio of *air* to fuel is 3.37 divided by 0.22, or 15.3 to one. In many books, this number is rounded off to 16; an air/fuel ratio of 16:1 is often quoted as "stoichiometric." The true value varies somewhat depending on the chemical composition of the gasoline.

Since it is often more convenient to speak in terms of fuel/air ratios, rather than air/fuel ratios, the above statements can be recast to say that a fuel/air ratio of 0.067 is considered stoichiometric, or chemically ideal, while F/A values greater than 0.067 are ''richer'' than stoichiometric, and (likewise) F/A ratios less than 0.067 are ''leaner'' than stoichiometric.

As it turns out, of course, you can get combustion to take place at fuel/air ratios far from the optimum. With suitable encouragement (in terms of spark, temperature, and pressure) it's possible to get combustion to occur within an engine at F/A ratios of from 0.055 to 0.125 (that is, A/F ratios from 8:1 to as high as 18:1). But by far the greatest amount of heat energy is released when there is neither an excess of fuel, nor an excess of air. When *every* fuel molecule and *every* oxygen molecule are able to combine together, with ''no one sitting out the dance,'' the maximum heat energy is released; there is no waste of fuel (nor of air). This is the point, of course, at which exhaust gas temperature (EGT) is maximum—the point of ''peak EGT.''

Airplanes, of course, have a manual mixture control on the panel, whereas cars do not. (In a car, the choke provides a rough equivalent of a mixture control, but even manual chokes have pretty much disappeared from automobile ''cockpits.'') The reason for this should be clear. Airplanes operate over a wide range of altitudes (whereas cars, for the most part, do not); and air density varies rather drastically with altitude. Yet, carburetors and fuel injectors cannot detect changes in air *density*. The result is that your carburetor (or injector) runs richer and richer the higher you go. If you were not able to lean your engine in some fashion at high altitudes, not only would you waste fuel, but your engine might even run roughly or quit due to the overrich mixture.

The effect of pilot-induced changes to F/A ratio on EGT indications is shown in Fig. 8-1. Pulling back the mixture control causes fuel flow to be reduced while airflow through the carburetor is held constant. (In a plane with a fixed-pitch prop, or a turbocharger for that matter, there will in actuality be some changes in airflow as the mixture is retarded; but for purposes of Fig. 8-1 we'll assume that the aircraft has a constant-speed prop, and that intake airflow is indeed constant.) The leanest F/A ratios occur on the left-hand side of the graph, while the richest conditions occur on the right-hand side. Exhaust temperature is plotted vertically. In this case, EGT is shown in terms of variation from peak-EGT; but you could just as easily plot ''absolute'' EGT values (1,200, 1,300, 1,400 degrees F, etc.) along the vertical scale at the top of the graph. The reason actual EGT values are not plotted is that they vary in magnitude from engine to engine, and from throttle setting to throttle setting, whereas variations from peak-

EGT are always about of the same magnitude. As plotted, the graph can be considered representative of a wide range of engines.

Notice that as the mixture is retarded, EGT rises in more-or-less straight-line fashion until, at around .067 F/A ratio, the curve peaks and starts back down. The peak-EGT point occurs, as we said, where the fuel/air ratio is such that there is neither an excess of fuel nor of air—all of the reactants are present in the chemically ideal amounts for combustion. Since the throttle setting in this graph is constant, the peak-EGT point represents the F/A mixture at which maximum heat is produced with respect to the amount of air being drawn into the engine. This is *not* the same F/A mixture, however, as that which gives the maximum *power* for the throttle setting selected. Maximum power occurs at about .076 F/A ratio—or 100 to 125 degrees F rich of peak. Why don't peak-EGT and "best power" coincide? The answer has to do with gas-expansion effects. At a F/A ratio slightly richer than peak, an excess of combustion gases (from incomplete burning of the excess fuel) is produced. These gases, like carbon monoxide, are capable of further oxidation (further burning) to yield a relatively small amount of extra heat energy, but are less dense and therefore take up more space than either their completely burned byproducts or the original fuel vapor. The extra gases produced through incomplete burning thus actually contribute more to piston movement than they would if burned completely to yield a modicum of extra heat. Piston movement, not heat *per se*, is what results in power; hence "best power" mixture occurs slightly rich of the peak-EGT fuel/air ratio. (If you want, you can think of "best power" as the *maximum-pressure* mixture point, and "peak EGT" as the *maximum-heat* mixture point.)

The peak-EGT point on the curve, then, can be seen as the mixture that gives the greatest amount of *total* combustion heat per unit of air (remember, airflow through the engine at a given throttle setting is constant in this example); while on the other hand, "best power" mixture is the F/A ratio giving the most *usable* heat (the most power) per unit of air. This is quite a different thing from getting the most power from a given unit of *fuel*, however. By definition, the best-power-per-unit-fuel mixture would also be the "best sfc" F/A ratio. (Specific fuel consumption, or sfc, is expressed in units of pounds of fuel per horsepower per hour.) The best-sfc point, also known as "best-economy mixture," occurs at or near 0.059 F/A ratio, or about 50 degrees F lean of peak-EGT. Why there? Because as you lean past peak, fuel consumption drops off faster than power production—up to a point. Eventually, as you continue leaning past "best

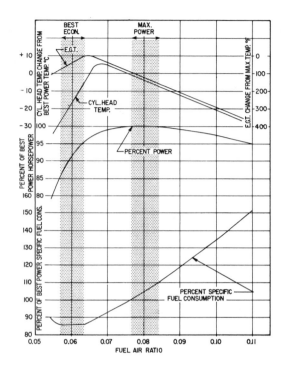

Fig. 8-1: Exhaust gas temperature versus fuel-air ratio. Although the actual temperature ttained at peak-EGT will vary somewhat from one engine to the next (indeed, from one cylinder to the next), the shape of the EGT curve and its relationship to F/A ratios stays the same for every engine. Note that ''best power'' and peak-EGT do not coincide. (See text for discussion.)

economy'' mixture, power production will drop off precipitously (i.e., your engine will cough and quit) even though some fuel is still flowing.

The effect of leaning on airspeed is worth considering. It should be obvious (even though it actually isn't to quite a few pilots) that maximum airspeed comes at *best-power* mixture—about 125 degrees (F) rich of peak EGT. Any leaning beyond this point will result in a reduction in airspeed. How much extra power does ''best power'' mixture produce over full-rich mixture? Generally speaking, about five to seven percent. Thus, if your plane develops 150 hp in cruise, you can expect to pick up about 10 horsepower by leaning from full-rich to best-power. How much airspeed you pick up depends on the individual airplane, but you can figure on a gain of roughly two and a half percent. So for example, if your plane normally cruises around 150 mph (full rich), leaning to peak will allow you to pick up just under 4 mph—a noticeable increase, just within the limits of measurability. Some airplanes will actually pick up more than two and a half percent, due to ram effects (the higher the airspeed, the more ram recovery and the better the volumetric efficiency of the engine). The point is, you can lean your engine to "best power" by reference to the airspeed indicator, if necessary (in calm air). The airspeed gain is just large enough to be perceptible to the average pilot flying the average-equipped airplane.

What happens to airspeed if you lean beyond best-power? According to the Cessna 182L owner's manual, a Skylane leaned to 75 degrees F rich of peak, rather than 125 F rich, will lose 1 mph in cruise—but will gain a full 10 percent more range (compared to best-power mixture). At 25 F rich of peak, the airspeed loss is 3 mph, but—if Cessna's handbook can be believed—the range increase *vis-a-vis* best-power is a full 20 percent! (Cessna does not allow leaning beyond 25 F rich of peak, in its 182L owner's manual, citing possible engine roughness due to premature onset of lean misfire in the leanest cylinders of the Continental O-470-R engine.)

Leaning to best-economy provides even more startling gains in range (assuming your engine can be leaned beyond peak without running roughly; many can't). Specific fuel consumption is typically 50 percent higher—yes, *50 percent*—at full-rich than at best-power mixture. The ratio of full-rich sfc to *best-economy* sfc is often 1.7 or better—almost a 2-to-1 ratio. In other words, if your range at full-rich mixture were 600 miles, your range at best-power mixture might be 900 miles; and your range at best-economy mixture might be 1,000 miles or more.

Fuel Schedules: Ideal vs. Real-World
From what we've said thus far, it is easy to imagine an "Ideal fuel-metering schedule" that would give maximum economy. After all, at takeoff—when maximum power is needed from the engine—there's a clear need for best-power mixture, while at very low power settings (approach, idle, taxi, etc.) the only real need is to keep the engine running, which means the ideal full-rich "idle" mixture would simply be the leanest fuel/air ratio that would support combustion. In between full power and idle, the carburetor could be designed so that with the mixture control "full rich," fuel is delivered at the best-economy F/A ratio. (Manual leaning will still be needed at high altitude, of course, to maintain the F/A ratio.) In this fashion, one could perhaps enjoy the best of all possible worlds: best-power on takeoff, best-economy in cruise, with only a small amount of manual leaning called for at altitude, to compensate for density effects.

The idealized fuel schedule just described encompasses fuel/air mixtures ranging roughly from 0.050 to 0.076. In stark contrast, the

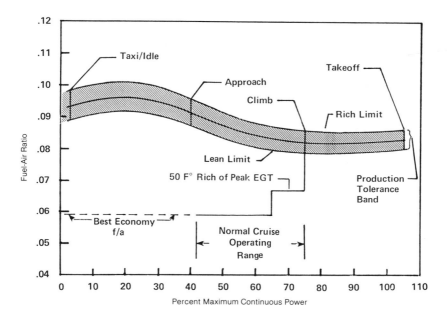

Figure 8-2: Mixture ratio versus percent power for Continental IO-520.

actual range of full-rich mixtures designed into the typical Lycoming or Continental engine spans F/A ratios from 0.080 to 0.105, with a 10-percent production tolerance being fairly typical. This is at least 50 percent richer than our postulated "best of all possible worlds" fuel-delivery schedule. (Fig. 8-2 shows the actual fuel schedule—mixture ratio vs. percent power—for a Continental IO-520 with TCM fuel injection.)

Why are aircraft engines set so rich? Mainly for cooling, and in some cases for detonation suppression. In an air-cooled engine, it is not unusual for 15 percent of a cylinder's cooling needs to be met by fuel. At high power settings, some excess fuel is needed to suppress detonation, as well as to keep the cylinder head within allowable temperature limits—i.e., under 460 to 500 degrees F. (Current FAA certification rules require the engine maker to set full-rich takeoff mixture a minimum of 10 percent richer than the rich-limit detonation-onset F/A ratio.) At low power settings, of course, cooling is not a particular problem; but transient throttle response is. At very low power (i.e., near idle), the mixture must be rich enough so that opening the throttle does not suddenly kill the engine. (When the throttle is opened, the engine experiences a sudden inrush of air; fuel delivery lags, especially in a fuel-injection system without an accelerator pump. Thus, a transient lean-mixture condition exists.) Near idle, the fuel metering system must be rich enough to give consistent cold-weather starting ability, and rich enough also for the engine to accelerate—and not die—when the throttle is opened. In practice, this calls for a fuel-air ratio of 0.100 (an air-fuel ratio of 10:1), or close to it.

There are other factors to consider as well, such as the fact that fluctuations in ambient pressure and temperature—i.e., day-to-day changes in density altitude—can throw off fuel/air ratios at sea level, with the result that the engine designer must allow a "fudge factor" for lean-day conditions. Also, not all of an engine's cylinders get the same exact amounts of fuel and air (see below), and some compensation on the rich side must be made for the whole engine, to protect the leanest cylinders from detonation, overheating, etc., under worst-case conditions.

In the absence of water cooling (which brings with it its own problems), air-cooled engines are necessarily rich-running monsters, at least under worst-case operating conditions. This is merely another way of saying that the well-informed pilot has much to gain, in the way of operating economy, from an understanding of how to lean his or her engine under better-than-worst-case conditions.

Mixture Maldistribution

Sadly, many general-aviation engines cannot be leaned to best-economy—or even to peak EGT without running roughly (even at reduced power settings). The reason is mixture maldistribution. Many small-plane-engine intake systems are so poorly tuned that lean misfire begins in the leanest cylinder(s) well before peak EGT is reached in the richest cylinder(s). Hence, arbitrary limits must be placed on leaning, for certain engines (witness Cessna's 25-degrees-rich-of-peak limit for the Continental-powered Skylane).

It's a fact of life that in all but the most carefully designed engines, some cylinders are richer than others. One reason for this is that the distance from the carburetor to the nearest cylinders is often (for large engines, such as the Continental O-470-R) half the distance from the carburetor to the farthest cylinders. Fuel droplets—traveling "uphill" against gravity (thanks to the updraft positioning of the carburetor)—are more likely to reach the near cylinders than the far ones; and the farthest-forward jugs run leanest (in some engines, anyway) as a result. Fuel injection is designed to overcome this limitation. But even fuel-injected engines experience significant inter-

Injector nozzles supply a controlled stream of fuel to the intake port of each cylinder. Gum deposits, if allowed to build up in the nozzle, will affect fuel flow, and EGT.

cylinder EGT spreads due to nozzle manufacturing tolerances, airflow anomalies, etc.

If you have a multi-cylinder EGT indicator system (a so-called ''exhaust analyzer''), it's a fairly simple matter to determine which cylinder is leanest. *Your leanest cylinder is the one that reaches peak EGT first as the mixture control is retarded.* As you pull the mixture control back in cruise, the EGT indication for each cylinder will initially rise, then fall, in accordance with the pattern shown in Fig. 8-1. Since peak EGT always occurs at or very close to a fuel/air ratio of 0.067, the cylinder that starts out nearest this value will reach peak EGT soonest—and by definition, that's your leanest cylinder.

Pilots often confuse the ''leanest cylinder'' with the ''hottest cylinder.'' Sometimes the two are the same; but very often, they are not. The cylinder with the hottest peak-EGT indication is not necessarily the leanest cylinder—it may be simply the cylinder receiving the largest total amount of combustants (fuel and air). Remember, it takes air *and* fuel to make fire, and generally speaking, the more you have of both (together), the hotter the flame. Also, the more compression there is during combustion, the higher the final temperature (which is another way of saying that your highest-peaking jug may well be the jug with the leak ring and valve leakage—the cylinder with the best compression). It's easily possible, if you think about it, for your *richest* cylinder to give the highest peak EGT value. The richest jug is the last one to peak. There's nothing that says it can't reach the *highest* peak. (However, there's nothing that says your richest jug is the one getting the *most fuel*. It could just as well be the one getting the *least air*.)

Again: Your leanest cylinder is the one that reaches peak EGT before all the others, as mixture is leaned. The actual magnitude of the peak is of little significance.

A corollary of what we've just discussed is that at any given fuel/air ratio, it's very likely that no two cylinders in your engine will show the same EGT indication. In fact, the difference between the highest cylinder and the lowest cylinder may well be 200 degrees Fahrenheit—or more. This is usually a direct reflection of the effects of mixture maldistribution. In other words, if your engine's intake risers were tuned and of the same length, if your injector nozzles were calibrated to high accuracy, etc., the inter-cylinder EGT spread would narrow, perhaps to zero. And as a result, you could lean your engine more aggressively. As stated above, one of the main limitations to leaning some engines to peak EGT or beyond is the premature onset of lean misfire in the leanest cylinders. Obviously, if every cylinder were equally "lean," you could lean your engine to a further extent before encountering lean misfire. And then the whole engine would just quit (instead of one cylinder)!

In some engines that have a large inter-cylinder EGT spread, it is possible to narrow the gap (and improve mixture distribution) by judicious use of carburetor heat. The improved vaporization of fuel made possible with carb heat aids in "homogenizing" the mixture in carbureted engines, and to a large extent mitigates the problem mentioned earlier of fuel droplets not reaching the distal cylinders (due to airflow problems, and gravity). Bear in mind, however, that carburetor heat—when applied fully—is enormously effective in heating the incoming fuel and air, and the use of full carb heat may cut volumetric efficiency (by reducing density flow). Also, carb heat involves the use of unfiltered air, which means TBO may suffer unless you are flying well above the "dust layer" (which extends to 10,000 feet AGL in certain parts of the country, in summertime). Partial carburetor heat generally does a better job of reducing EGT spread than full-strength carburetor heat; however, the effect varies from plane to plane. As mentioned above, some EGT variations are attributable to differences in ring or valve leakage (compression) from cylinder to cylinder. If this is the dominant problem in your engine, the use of carb heat will not tend to minimize inter-cylinder EGT spread.

Any time you use carburetor heat, by the way, remember that the incoming air—because it is hotter—is less dense (contains less oxygen per cubic foot) and therefore makes the effective fuel-air ratio *richer*. (Check this with your EGT. The EGT should go down as you apply carb heat.) Accordingly, you should *re-lean* any time carb heat is used.

Material Limitations

For most high-output aircraft engines, the principal limitations to aggressive leaning are material-related. The material properties of iron alloys deteriorate rapidly as the temperature increases above 1,200 degrees Fahrenheit. As tensile strength and hardness go down, wear of moving parts goes up. Also, corrosive attack (always a problem in components exposed to exhaust gases, which are rich in water vapor, acids, and ozone) is accelerated by high temperatures.

Then there is the problem of "creep," which is the tendency of metal alloys to become plastic and flow (or stretch) at very high temperatures. Generally speaking, the creep limit of most austenitic steels is encountered at around 1,400 degrees (it varies for different alloys). Creep properties are given in units of psi loading for a given percentage growth per 1,000 hours. (For low-grade carbon steels, a typical value might be 7,800 psi for one-percent growth in 1,000 hours.) With the proper engineering handbooks, it is possible to calculate with a pocket calculator how much a turbine blade, or an exhaust valve, will "grow" in 1,000 or 2,000 hours of operation at a given temperature. Just this type of calculation is often used to establish hot-section or major-overhaul TBOs.

The exhaust valves used in modern aircraft engines (as explained in Chapter 1) are usually fabricated not from ordinary steel alloys, but from nickel-based "superalloys" such as Nimonic 80A. These high-nickel alloys provide excellent creep resistance to temperatures of 1,700 F or more. Unfortunately, not every component in an aircraft engine's top end is made of such alloys. (Exhaust stacks are commonly made of 321 stainless, for example, which loses 90 percent of its room-temperature tensile strength at 1,650 F—a temperature easily attainable in many light plane exhaust systems.) Valve guides, in order to provide good wear characteristics, must generally be softer than the valve-stem material designed to run inside them; hence aluminum-bronze and cast iron guides are often specified for aircraft engines.

The point is, many of the metals used in your engine, for one reason or another, are not capable of holding up well under extreme-high-temperature operating conditions, and there's little that can be done about it except to monitor EGT (or TIT) closely, and keep temperatures in check. Fortunately, quite a few of aviation's low-output engines are incapable of producing extremely high EGTs in normal operation. For example, you would be hard-pressed to see EGTs of 1,600 F in a Lycoming O-235-N2C or Continental O-200-A with both mags working, timing adjusted to factory specs, etc. (In fact, leaned to peak EGT in cruise flight, you'd be hard-pressed to see 1,400 F in either of those engines.) Even some high-compression engines, such as the 200-hp Lycoming IO-360-A used in the Mooney 201, are fairly cool-running with regard to EGT. This is not unexpected, since the heating effects of the high compression ratio are usually offset by the (equally high) expansion ratio, which allows exhaust gases to cool before passing the EGT probe. (Here we see yet another subtlety of exhaust analysis, which is that compression ratio and EGT are not necessarily related—something that was first noted by Alfred Buchi in 1909.)

Nevertheless, aggressive leaning *can* produce valve temperatures (and exhaust-stack temperatures) exceeding 1,400 F in many of today's light plane engines, particularly turbocharged engines. To some extent, Lycoming engines are better protected from heat damage than Continental engines, in the sense that Lycoming exhaust valves are sodium-filled and operate cooler, by a significant margin, than solid-stemmed Continental exhaust valves. Even so, it is a good idea to have a regard for exhaust-temperature limits when leaning an engine—any engine. Not every EGT is calibrated in actual degrees Fahrenheit. Calibrating an EGT system is not difficult, however (Alcor makes an excellent calibrator system; see if your shop has one); and in the long run it will pay dividends to put a red line on your EGT gauge corresponding to 1,650 F (or some lower value).

One of the unavoidable trade-offs of flying is that the more aggressively you lean your engine, the hotter your exhaust components are going to run and the less life they'll have. This is more of a limitation for some engines than for others: Obviously, a 375-hp Continental GTSIO-520-M (with solid-stemmed valves) operating at 30 inches of manifold pressure in cruise will not tolerate the same degree of leaning as a 180-hp Lycoming O-360-A (with sodium-cooled valves) pulling 22 inches in cruise. The GTSIO-520 *can be operated* the same way, but you'll pay for it in top overhauls (valve jobs) later.

What Does an EGT Probe Measure?

You can't intelligently interpret any instrument's indications unless you understand how that instrument senses what it senses and displays what it displays. This is as true for EGT systems as it is for airspeed indicators or altimeters.

EGT systems (as explained in Chapter 2) come in single- and multi-probe varieties. The "probe"—a chromel/alumel thermocouple, usually—may be inserted at any of a number of locations in any exhaust pipe of the engine. (In single-probe systems, the sender may be placed near the outlet of one exhaust port; or it may be placed at the nexus of a 'Y' connection, in which case the instrument indication is an "average" of the EGTs of two or more cylinders.) The location of the probe will have a profound effect on the indications seen in the cockpit; you can change your EGT magically just by changing the location of the probe.

What, exactly, does the EGT probe sense? Obviously it senses exhaust gas temperature, but in a four-stroke-cycle engine, the exhaust valve in any given cylinder spends approximately two-thirds of its time in the closed position. Thus, the EGT probe spends two-thirds of its time sensing—what, exactly? Lack of exhaust gas temperature? Radiant heat from the cylinder head? Nothing at all?

In a multi-probe system, where each exhaust riser has its own "dedicated" probe, you can picture the probe sitting idle two-thirds of the time, then being blasted with a very-high-temperature, high-pressure "pulse" of exhaust gas (which itself is in the process of expanding and cooling). The probe thus is exposed to a regularly recurring cycle of events. The only reason you don't see the needle flicker rapidly on the cockpit gauge is that the combustion cycle is of overall shorter duration than the response time of the probe and needle. The probe itself, after all, has a certain amount of mass, and therefore thermal inertia; and the transfer of heat from a gas to a metal takes time, etc. The point is that what you see on the cockpit gauge is nothing but a rough moving average of the temperature of a certain time-averaged event. It is *not* an indication of instantaneous exhaust gas temperature.

If the object is simply to have something to lean the engine by, why not just put an automotive-type oxygen sensor in the exhaust pipe and be done with it? (The oxygen sensors used in modern auto emissions systems actually sense the oxidation potential of the exhaust and therefore permit a direct determination of fuel/air ratio, without dynamic variations to find a "peak" in the mixture curve.) The answer is that an EGT system gives much more information about combustion than an oxygen sensor ever could. The EGT probe is the engine-maker's equivalent of a medical thermometer—it senses combustion directly (more or less), and therefore gives the operator valuable insights into the combustion process that aren't available by any other means.

Consider the normal combustion cycle: The intake valve opens just as the exhaust valve is about to shut (both are actually open at the same time, briefly; this is known as "valve overlap") and the fuel-air charge is drawn into the cylinder on the piston downstroke. On the compression upstroke, of course, both valves are closed. Ignition occurs at about 20 degrees of crankshaft travel prior to the piston's reaching top center. A flame front spreads outward from each spark plug, and peak combustion pressure is reached very near the crest of piston up-travel. (Peak temperatures reached in combustion are 4,000 F or more.) Combustion is complete just after the piston starts down, and at around 50 degrees of crank travel before piston bottoming, the exhaust valve opens. Only then does the EGT probe get a "peek" at the leftovers of the combustion process.

Obviously, anything that postpones or accelerates combustion can and will have an impact on EGT, given the sequence of events just described. For example, if ignition timing is shifted early (advanced), so that the spark plugs fire at 30 degrees before top center on the piston upstroke instead of 20 degrees BTC, combustion will go to completion slightly earlier in the cycle than it normally would. As a result, the exhaust gas has more time to cool—more heat is dissipated to the cylinder head—before the EGT probe "sees" what's left. Indicated EGT thus goes down.

Likewise, if ignition occurs late, combustion occurs late and exhaust gases are released earlier in the cooldown cycle than before; hence EGT goes up. This also happens if one spark plug stops firing. Combustion duration is shorter with two flame fronts than with one. Eliminate one spark plug from the combustion event, and peak combustion temperature is reached later than it ordinarily would be; EGT consequently goes up. (Take one magneto off-line, and EGT will go up about 200 degrees for *all* of an engine's cylinders.)

What happens to EGT when detonation (combustion knock) is encountered? Detonation occurs *after* spark-plug firing, when the unburned portion of the fuel-air charge ahead of the flame front autoignites and explodes all at once (producing the familiar ping or knocking sound). Detonation can be caused by many things: poor-quality gasoline, intake air that's too hot, improperly advanced timing, operating at excessive brake mean effective pressures without proper cooling, etc. Regardless of the cause, however, the effect on EGT is the same: Combustion takes place early, all at once, and heat is released (violently) to the cylinder head and piston. CHT goes up, while EGT goes down. (Notice that CHT and EGT are not always positively correlated. Quite often, as here, they are in fact *negatively* correlated.)

The mere fact that EGT goes down (as seen in the cockpit) does not necessarily mean detonation is occuring, however, or ignition has drifted early. It may simply mean you've burned a valve, or a bit of carbon is preventing a valve from closing completely, or piston blowby has increased—any of which would reduce compression (and thereby lower the peak combustion temperature). If EGT goes down in *all* of your cylinders at once, of course, it is unlikely that low compression is to blame. (It's hard to imagine a situation in which every piston suddenly developed excessive blowby, every valve got carboned up, etc.) This is where a multi-probe system proves its worth—*if* the pilot knows how to interpret it.

Remember, exhaust gas temperature is a reflection of many things: cylinder compression, fuel/air ratio, total fuel and air admitted to the cylinder, combustion duration, valve timing, expansion ratio (the factor by which the exhaust gas is allowed to expand before it exits the combustion chamber), and burn time, to name a few. One reason EGT initially goes up as you lean your mixture is that—all other things being equal—lean mixtures take longer to burn than rich mixtures, and the exhaust gas is hotter when it first leaves the cylinder when the fuel/air ratio is lean.

Many pilot's operating handbooks refer to the EGT gauge as an "economy mixture indicator." It is really much more than that. The EGT is a microscope trained on the combustion process, and if you know how to use it, you can "see" a great deal indeed.

CHAPTER NINE
DESCENT, APPROACH,
& SHUTDOWN

Descent, Approach and Shutdown

The transition from cruise to tiedown is a critical phase of operation, yet deceptively simple from the novice pilot's point of view. In a trainer aircraft with a stout, reliable powerplant unencumbered by wastegates, controllable propellers, prop reduction gearing, dynamic counterweights, etc., it is acceptable simply to yank the throttle back and head downward. Unfortunately, what's good enough for the Cessna 150 is not good enough for the Navajo Chieftain. Getting the high-performance engine back to earth without doing damage calls for a bit more expertise.

Shock Cooling

Mechanical stress and thermal stress are the two greatest killers of aircraft engines, and a rapid letdown presents ample opportunities for both. Mechanical stress arises from vibrational modes and transient propeller and crankshaft loadings that can occur in the descent (see discussion below). Thermal stress, of course, occurs through shock cooling. Shock-cooling has been the subject of some debate in recent years, with a few people even going so far as to say it doesn't exist at all. (This is, of course, tantamount to saying that cooling fins on cylinders do not perform a useful function.) Any operator of jump planes or glider towplanes will readily attest to the destructiveness of shock cooling. Granted, it is a poorly understood phenomenon. But to say that shock cooling is poorly understood is not to say that it doesn't exist. It is as real as tooth decay.

Continental engineers Bernard Rezy, Kenneth Stuckas, J. Ronald Tucker, and Jay E. Meyers write about the subject in SAE Technical Paper No. 790605 (Society of Automotive Engineers, 1979), stating: ''[A phenomenon that is] difficult to quantify is the reduction in cylinder head durability due to low cycle thermal stress superimposed upon the high cycle mechanical stress of combustion pressure. This low cycle thermal stress is caused by the normal aircraft oprating cycle—high cylinder head temperatures during takeoff and climb modes of operation and low temperatures during high-speed descent modes. Leaning to best-economy in descent at low powers results in cylinder head temperatures which are lower than at best-power or even full-rich fuel flow. *This overcooling effect can have an impact on cylinder head life."* (Emphasis added.)

Right away we see one possible clue to minimizing shock cooling's effects: Do not descend with the mixture leaned to best-economy at low power. Keep some power on the engine—and re-lean as necessary to keep EGTs high (or rather, near cruise values).

The potential for shock cooling of an aircraft engine (especially an *air-cooled* aircraft engine) should be obvious. Airplane engines routinely operate at mountaintop altitudes, where the air is frigid (or worse). In cruise, an air-cooled aircraft engine, operating at or near full throttle, relies upon massive quantities of that frigid air for cooling. In a rapid descent, however, power is reduced dramatically (so heat output is suddenly low), yet at the same time, airflow through the cowling is vastly increased—the perfect setup for thermal stress.

What kinds of mayhem does shock cooling visit upon an air-cooled powerplant? The main damage is to the hottest components in the engine—primarily cylinder heads, valve seats, and valves. Aluminum cylinder heads shrink twice as much, per unit of temperature loss, as steel cylinder barrels (or steel valve inserts). Thus, valve seats can be distorted—or even pop out—as the head cools, and the area around the interface of barrel and head casting can develop stresses that lead to later cracking. Valves can warp from the sudden temperature change (if it is sudden enough), or even stick in their guides. According to Lycoming's Ken Johnson, writing in the *Avco Lycoming Flyer* (May, 1982): "Engineering tests have demonstrated that valves will stick when a large amount of very cold air is directed over an engine which has been quickly throttled back after operating at normal running temperatures." When a valve sticks in anything but the full-open position, the pushrod and lifter have nowhere to go, and something's got to give; usually it's the pushrod (which takes a permanent bend), although the lifter can also crack. (When a valve sticks in the full-open position, it contacts the piston and sends metal through the engine—and through the turbocharger, if there is one.)

Perhaps the most common type of thermal stress-relief damage is cylinder head cracking in the area of the spark plug holes or the valve inserts. These stress cracks—which radiate outward from valve seat or plug bosses, usually in the direction of the nearest adjacent boss or opening—are often not discovered until the engine is disassembled for major overhaul and the cylinders are dye-penetrant-inspected. (At that point, the cylinders are often scrapped, although in many cases they are repairable.) Nor is this a form of damage limited to large-displacement, high-output engines. Lycoming O-320 and O-360 cylinder heads often show this type of damage, particularly in high-time trainer or club aircraft. "Any O-320 cylinder with more than 3,000 hours on it stands a good chance of having cracks in it," one experienced engine overhauler told us flatly.

The insidious thing about shock cooling is that you may not be able to monitor it adequately with CHT indications, owing to the way manufacturers locate CHT probes. By regulation, airframe manufacturers are required to locate an airplane's cylinder head temperature sensing probe on the overall hottest cylinder (as determined by flight test). This location can vary even within a given aircraft model series, depending on the location of cooling baffles, engine accessories, etc. from year to year. But in any case, the cylinder that your probe is installed on is apt to be high-indicating all the time (relative to your other cylinders—which of course do not *have* a CHT probe, unless you've bought an aftermarket kit), precisely because it has been placed there on the basis of that cylinder being hottest in all phases of flight. What you're looking at in the cockpit, when you watch the CHT gauge, is your hottest cylinder. You have no idea what your *coldest* cylinder is doing.

Prevention, in any event, is easy (if you can manage it): Reduce power gradually for descent. "It is poor technique to 'chop' the power from cruise or higher settings to idle and then start a rapid letdown, which develops excessive cooling airflow over the engine," explains Lycoming's Ken Johnson. "It is always best to reduce power in increments, so that engine temperature changes will occur gradually." Of course, in a descent of 500 feet per minute, the throttle *must* be pulled back at a rate of one inch of manifold pressure every 120 seconds just to keep a constant power setting! (Ambient pressure increases by one inch of mercury with every 1,000 foot drop in altitude.) In fact, in an airplane with significant MP ram rise—such as a turbocharged Cessna twin—the mere picking up of 20 knots (IAS) in the initial descent can add an inch of manifold pressure to the MP gauge by itself, before significant altitude has even been lost. This brings up the question of whether a descent ought not to be initiated with elevator trim rather than with throttle. The answer is probably no, if you are in a "hot" airplane (Bonanza, Baron, 310, etc.), as airspeed can quickly build to redline—or at least into the yellow arc—if power is left constant during the descent. Going to lower altitudes, your chances of encountering turbulence are, of course, greatly increased; and you want to be out of the yellow arc on the airspeed indicator when you hit unstable air (turbulence altitudes). So it is definitely advisable, from a turbulence-penetration standpoint, to keep a reasonable Mach number in the descent. And that means reducing power.

How quickly should you reduce power? There is no one correct way to select a power setting—or a power-reduction gradient—for letdown. The answer is simply to keep enough power on the engine (and keep the mixture just lean enough) to maintain stable CHT and oil temperatures, or at least allow a gradual cooldown, as the descent progresses. This translates into various techniques for various airplanes. Many pilots feel that a rate of power reduction of two inches initially, followed by one inch of manifold pressure per three or four minutes (or as necessary to allow a descent at an indicated airspeed no greater than the cruise IAS) is appropriate under most conditions. This assumes that the airplane has no aerodynamic features (speed brakes, for example) that allow a rapid descent without a significant power reduction. In many airplanes, the "flaps-10" position can be used at indicated airspeeds above normal cruise speed, so that a descent can be initiated simply by lowering the first increment of flaps. In others, the landing gear may be lowered at fairly high speeds. (In still others, the gear must be lowered at low speeds, but once down and locked, can be flown at much higher airspeeds; consult your owner's manual for details.) If a special drag-inducing feature is available, and if it is practical to do so, by all means use *it*—rather than a power reduction—to initiate your descent. As the letdown progresses, power can be reduced at a gradual rate (e.g., two to three inches every minute) to prevent thermal stresses from damaging top-end components.

Some pilots, believe it or not, are in the habit of using cowl flaps or retractable landing lights as speed brakes to initiate letdowns—which is definitely a bad idea. Opening the cowl flaps to increase drag will only accelerate cold-shocking; if anything, you want to close the cowl flaps a notch as you bleed the power off. Likewise, landing lights (retractable variety) may not be able to take the stress of repeated deployment at high airspeed, so if you don't want to end up landing in the dark someday, don't use retractable lights as speed brakes. (In the Cessna 310, this is an especially dismal idea, since the main-fuel-tank transfer pumps are wired through the landing-light-motor circuit breakers. Blow the CBs and you just may finish your approach in the dark, with two dead engines.)

Should you perhaps sideslip the airplane to increase drag in the descent (thereby allowing more power to be used, to stave off overcooling)? Probably not. Huge bending loads are imposed on the prop and crankshaft whenever you sideslip an airplane at high speed or at high power settings. Reject the idea as ludicrous.

Gradual power reductions are fine to talk about, but of course in the real world, ATC (air traffic control) may not "allow" you to descend at an engine-saving rate. (Actually, the *pilot* "allows" the controller to control the airplane, and doesn't—strictly speaking—have to accept a descent clearance that might damage an engine; but it's wise not to let such situations develop when they can be avoided. When possible, ask for descent clearances far enough out that you can plan a minimum-rate letdown—which, for IFR purposes, generally means 500 fpm.) Here again, there are no easy answers: If a controller tells you to lose altitude at a rate of 1,000 fpm (or more), and you don't have speed brakes, you're stuck, basically. About the best you can do, if you're still far from the destination, is reduce manifold pressure and rpm as needed, *close the cowl flaps all the way*, and start down. (Re-lean to peak EGT in the descent.) If you're near the destination, naturally, you can consider dropping the initial increment of flaps and/or lowering the gear (depending on which V-speed limit you can reach most quickly) before deciding on a final power-lever setting.

Incidentally, in your zeal to get the gear and flaps down, do not (unless you *know* it's safe) simultaneously activate electrically driven flaps *and* electrically driven landing gear. If you're already using significant electrical power to operate radios, autopilots, etc., the sudden load incurred by simultaneous usage of gear and flap motors is apt to blow your alternator circuit breaker, perhaps with the gear halfway down. (Don't count on your battery to get the gear all the way down.) Remember, landing gear motors often have a peak draw of 25 amperes or more; flap motors, 15 amps; pitot heat, 10 amps; landing lights, 10 to 20 amps. Add radios to the foregoing bunch, and you've "maxed out" the typical 60-amp alternator. (Again, don't count on your battery to make up the difference.) *Operate one major accessory at a time.* It's just good common sense.

Momentary power reductions on the order of six or eight inches of manifold pressure (with an appropriate decrease in prop rpm) for purposes of reaching gear-lowering speed, or flap-lowering speed, should not be considered harmful to the engine *if* power is again restored to the engine as soon as the gear is down. (After all, while you're throttling back, airspeed is falling off quickly, and cooling airflow through the engine compartment is cut commensurately—which is good.) What you want to avoid is "chopping" power, or moving the power levers back most of the way—and leaving the power off while airspeed builds. Sustained high airspeed following a sudden, permanent reduction in power is what causes thermal-stress-related damage.

Also bear in mind that the degree to which power can be safely reduced for letdown is in no small way related to the amount of power you were using in cruise to begin with. Obviously, the worst possible case is to go from takeoff power (or high cruise power with CHT and oil temperature at or near redline), to no power at all. A sudden reduction of power from 55 percent to 35 percent is, by comparison, not much of a reduction. So don't be *too* squeamish about reducing power. We all have to do it sometime.

Counterweight Detuning

Except in an emergency situation, an aircraft engine's power controls should never be moved rapidly. (An exception is made, of course, for aerobatic airplanes; but aerobatic engine TBOs are measured in three digits.) Any power reduction for purposes of letting down should be made smoothly, to avoid possible detuning of crankshaft counterweights.

Crankshaft dynamic counterweights and used on some engine models—mostly six and eight-cylinder engines, but some four-cylinder Lycomings as well—to dampen out potentially harmful torsional (twisting) vibrations of the crankshaft. As explained in Chapter 1, these counterweights take the form of swinging pendulums mounted on blade-like extensions of the crank cheeks. The counterweights are held to the hanger blades by means of roller (pins) which ride in precision-machined bushings in the weights and blades. The size of the rollers, and the ID (inside diameter) of the bushings, determine the harmonic of the counterweight. This way the mass of opposing counterweights can be kept the same (for engine balancing purposes) even though they may be of differing frequency ranges.

In normal operation, the engine crankshaft operates at one rpm for extended periods of time, and the counterweights assume one position. If the operating rpm coincides with a vibrational mode (not usually the case in cruise), the counterweights oscillate at a particular frequency to counteract the crankshaft vibrations. During the *transition* from one rpm to another, however, the counterweights may not "track" the crankshaft closely, particularly if the rpm change is rapid. In fact, if it is rapid enough, the counterweights may slam back and forth in their mounts, or "detune" (even accentuating, rather than dampening, normal vibrations). Brinelling, cracking, or deformation of counterweight bushings may result. Or worse, stresses may develop which lead to later crankshaft breakage. This, in a nutshell, is why throttle reductions should be carried out smoothly and slowly during the descent phase of any flight.

Lycoming Service Bulletin No. 245 (which applies to *all* Lycoming engines that use counterweights), explains the problem this way: "Detuning the counterweight system of the engine occurs when the engine operates outside of its normal range and by abrupt throttle change. When this happens, the dynamic counterweights cannot follow the spectrum of frequencies for which they were designed and rapid and severe damage to the counterweights, rollers, and bushings results; culminating in engine failure." Since the design of Continental counterweights does not differ significantly from Lycoming's, this statement can be considered valid also for Continental engines that use counterweights—which includes just about every six-cylinder model that Continental makes.

Lycoming Service Bulletin No. 245 delineates four operating conditions that can cause the counterweight system to detune: (1) Rapid throttle movement. (2) High engine speed in conjunction with low manifold pressure. (3) Excessive speed and power, or overboosting. (4) Propeller feathering. To these, add one more: Sudden prop stoppage for any reason (gear-up landing, collision with snowbank).

On the subject of "high engine speed and low manifold pressure," Lycoming S.B. 245 merely explains: "Any operating procedure involving high speed an low manifold pressure (under 15 inches) such as might be the case during a 'power-off' descent can cause detuning of the counterweight system." This is particularly true for geared (prop-reduction type) engines, whose huge propellers can "overdrive" the crankshaft quite easily (by flywheel effect, and by windmilling) whenever the throttle is "chopped" or brought back rapidly. In a plane with geared engines, therefore, it is considered acceptable—within limits—to reduce *rpm* before reducing *throttle* to initiate a descent (which is of course the opposite of normal constant-speed prop management; usually you decrease manifold pressure with the throttle *first*, before reducing rpm).

Induction Ice

Induction ice can take two forms: carburetor ice or impact ice, the former affecting only carburetor-equipped aircraft (obviously), with the latter affecting, potentially at least, both carbureted and fuel-injected aircraft. Carburetor ice is any ice buildup in the venturi or throttle butterfly area of a carburetor (usually a float carburetor, although it can and does occur occasionally in pressure carburetors as well). Impact ice, on the other hand, refers to ice buildup at the air filter, air scoop inlet, or at sharp bends in the air scoop plumbing—or even at the throttle plate—of a fuel-injected engine installation. Impact ice builds in the same fashion, and under pretty much the same conditions, as airframe ice, which means you have to be inside a cloud—or else flying through wet snow or rain, at or near the freezing level—to get it. By contrast, carburetor ice can form in VFR conditions, far from the freezing level, with or without precipitation.

It's not necessary to be in the descent phase of a flight to encounter induction ice, of course (impact ice can form just as easily in cruise as in a letdown), but in practical terms, most carburetor ice is accumulated with the throttle closed, and most high-performance aircraft (which are fuel-injected and subject only to impact-type induction ice) do their cruising above the weather, ATC permitting. So in practical terms, you're *more likely* to encounter induction ice on the way down than any other time, especially if you fly a carburetor-equipped aircraft.

Just how much of a menace is carburetor ice? The National Transportation Safety Board found some years ago (in 1972) that in a typical five-year period some 360 accidents could be found in which carburetor ice was either the cause or a contributing factor. Only 40 fatalities were counted among the 360 accidents, however, and in as many as half the accidents, the only injuries were to the airplane. One of the interesting findings in NTSB's 1972 study of carb ice is that by far the largest number of carb-ice accidents involved Cessna 150s (47 in all). Other planes with an apparently high involvement in carb-ice episodes are the Cessna 182 and the Piper Apache.

Carburetor designs, of course, haven't changed much over the years, and neither have the basic facts surrounding carb ice: Carburetor ice can form at outside air temperatures as high as 90 degrees Fahrenheit, if humidity is high (i.e., high enough to cause some reduction in flight visibility—obviously you can't get carb ice on a bone-dry CAVU day). The water vapor in the air condenses as ice at the throttle plate because of the large temperature drop that occurs when gasoline evaporates in the throttle barrel, and also due to the "delta-T" that accompanies the pressure drop in a carburetor venturi. Throttle ice doesn't form in a fuel-injector system because fuel evaporation takes place at the intake ports (which are always toasty-warm), and the delta-P (pressure drop) across the throttle butterfly is not usually large enough to encourage ice formation. Like airframe ice, carburetor ice has a lower temperature limit: At OATs much below freezing, carb ice doesn't form, simply because the water vapor in the air is already frozen into hard crystals by the time it enters the venturi, and the crystals don't want to stick. (Unless, that is, you apply partial carburetor heat. Partial carb heat can melt incoming ice crystals and cause them to refreeze on the throttle plate. If you use carb heat, use *full* carb heat, unless you are in clear air and using partial heat for purposes of reducing mixture maldistribution—see Chapter 8. Also, expect a momentary power loss—maybe even engine stoppage—as any ice that is present "slushes off" and goes from the carburetor into the engine.)

The insidious thing about carburetor ice from the pilot's standpoint is that it is almost impossible to detect from the cockpit until it's too late (unless, of course, you have a special aftermarket carburetor ice detector—and even there, some operator skill is called for). The first indication of carburetor ice is usually a falling off of rpm or manifold pressure, or failure of the engine to run smoothly or develop poer when the throttle is advanced—if, indeed, the throttle *can be* advanced. Sometimes, the first indication of carb ice is a frozen throttle (literally). Hard-packed, snow-cone-like ice will often accumulate in the venturi to the point that the butterfly can't move. Then you're in real trouble.

Why not just apply full carburetor heat prophylactically, and leave it on, whenever OAT and humidity conditions seem to warrant it? During an extended descent, this isn't at all a bad idea. The coolest carburetor temperatures occur with the throttle butterfly closed (which creates an enormous delta-P and also cuts heat from the engine, since the engine is no longer putting out power); and during a descent, when you don't need much power anyway, you can accept the power loss that comes with use of carburetor heat. Under any other circumstances, however, the indiscriminate, fulltime use of carb heat (for prevention purposes) is generally not acceptable. The hotter intake air, because it is lower in density, causes a power loss on the order of 15 percent; mixture is affected; and according to Lycoming's O-360 Operator's Manual (p. 3-10), "High charge temperatures also favor detonation and preignition, both of which are to be avoided if normal service life is to be expected from the engine."

In short: Apply carb heat all the way or not at all; and if ice is present, expect a significant power disruption as the ice melts and sloughs off into the cylinders. (Power will return again.) Do go ahead and apply it preemptively if you are anticipating a long, slow letdown in hazy or cloudy conditions. Otherwise, if you're not in a prolonged descent, monitor power gauges and apply carb heat any time you suspect carburetor ice (i.e., if there is a gradual, unexpected reduction in manifold pressure or rpm, or the engine won't accelerate normally). Re-lean as necessary following application of heat. And stow the heat knob before initiating a go-around.

The remedy for impact ice, of course, is alternate air. Here again, pilot actions are limited—for the most part—to watching for the telltale signs of ice (decreasing manifold pressure), and acting after-the-fact, rather than taking preventive measures. (Some planes have air-scoop heaters or boots surrounding the air-intake duct, if the duct is in the wing.) When impact ice is suspected, the proper action is to pull the alternate-air knob on the panel—if one is available. In some airplanes, the alternate-air door is automatic (spring-loaded and actuated by engine suction), with no knob at all on the panel, in which case the pilot can take *no* action and theoretically need never worry about impact ice. The problem, however, with alternate air—automatic *or* manual—is that, like carburetor heat, it can have a noticeable effect on engine power output. In fact, some aircraft owner's manuals (like the one for the Cessna T303 Crusader) warn that as many as six inches of manifold pressure may be lost if alternate air is selected in cruise. This is because of the loss of ram air recovery (and also because alternate air is usually taken from a hot area inside the engine compartment). Of course, if you've already lost six inches of manifold pressure in cruise due to impact air blocking your air filter, selecting alternate air will have little or no effect (or, hopefully, an augmenting effect) on manifold pressure. If you do not have ice, though, and you do not see a MP drop when selecting alternate air in flight, you can suspect a leaky or stuck-open alternate air door, which should be looked at by a mechanic at once.

Finally, remember that alternate air and carburetor heat air are both unfiltered sources (i.e., the air filter is bypassed), allowing airborne dirt to enter the engine directly. The use of alternate air or carb heat should thus be confined to periods when it is actually needed; once you're out of an icing situation, stow the heat (or alt air) knob immediately.

Approach and Go-Around

Power management on the approach is limited, for the most part, to selecting a final power setting that will get you to the runway environment while also choosing full-rich mixture, full-high prop rpm, wastegate open (if applicable), and boost pump on (again if applicable). The prop pitch control, if one is present, should be moved forward very gradually unless the airplane is at an airspeed slow enough to preclude surging. In many airplanes, a sudden movement of the prop control towards the panel will, at airspeeds above 100 knots, result in an unpredictable (and frequently alarming) surge in rpm to redline—or beyond. This can be avoided by slowing down to approach speed *before* moving the prop to fine pitch.

The final power setting on the approach will vary from plane to plane (and from approach to approach, naturally). For a Cessna 150 making a no-flap landing, idle throttle is, of course, entirely appropriate. In larger (heavier) airplanes, landings are routinely made with some power all the way to touchdown, or at least into ground effect. In a Cessna 310, for example, one typically uses 17 inches of manifold pressure all the way down final approach, retarding throttles in the flare.

Cowl flaps should be adjusted with some regard for prevailing OAT, CHT, and indicated airspeed. To state flatly that all approaches should be flown with cowl flaps closed (or open) would be misleading. On a cold day in Fairbanks, it may be necessary to keep the cowl flaps closed from takeoff to touchdown. Conversely, on a summer day in Albuquerque, you'll probably want the cowl flaps open on final approach, especially if you've been doing any slow-flight maneuvering (due to ATC vectoring, for example) before initiating the approach. Remember that during slow flight in the pattern (or in the terminal radar service environment), with airflow through the cowl reduced to a trickle of what it was in cruise or on let-down, significant heat buildup can occur even at low power settings. After a minute or two of holding at one altitude with the gear and flaps down, it's almost always a good idea to crack the cowl flaps open, even on a fairly cool day. If it should be necessary to subsequently "chop" power, you can then close the cowl flaps once more.

Use common sense with cowl flaps. The engine compartment and everything in it represents a fairly large heat sink, with "thermal inertia" properties not unlike those of a house or a car. (It can get heat-soaked or cold-soaked, but significant temperature changes do require time.) The idea is not to keep temperature fluctuations from happening—since that's clearly impossible—but to keep CHT and oil-temp changes *gradual* and well-dampened. Airspeed (which has an effect analogous to "wind chill") and cowl flaps are your number one and number two tools for modulating FWF (firewall-forward) temperature swings. Remember that any time you slow the airplane down, cooling is cut just as surely as if you had closed the cowl flaps. Try to form a mental picture of temperature *trends* as they are occurring, ahead of the firewall; and use all the tools at your command, in concert, to effect the conditions you want. This is something that comes with experience.

Should mixture be kept leaned during final approach? It can be, but unless you're flying in and out of a very-high-density-altitude airport, it is probably not a good routine procedure. Most of the powerplant-related items on your pre-landing checklist (mixture rich, prop control forward, boost pump on) are designed to allow safe transition to go-around mode should the approach need to be aborted. During a go-around, you want the maximum amount of power to be available *immediately*. When you have to avoid hitting another aircraft, seconds count (and percentages of power count). With the mixture leaned, your engine may detonate or falter, or fail to develop power, during a sudden application of throttle. Likewise, with the prop control in a low-rpm setting, the engine may not accelerate properly, and in any case it will not put out the maximum available horsepower for climb. Deviate from the pre-landing checklist at your own risk.

And by the way, be as brisk with the throttle as conditions warrant during a go-around or any other emergency procedure. In a landing abort, it's your life—not the engine—you're trying to save *first*.

Shutdown

The period from touchdown to shutdown is a critical one for many engines. Turbocharged engines in particular need a cooldown period of four to eight minutes (running at idle on the ground) before ceasing operation, to prevent turbo coking. After landing, a turbocharger rotor is usually still at 1,000 degrees F or more. Practically the only thing cooling the rotor and shaft (other than black-body radiation) is the slow flow of engine oil through the bearings of the turbo. At temperatures above 600 F, ordinary mineral oil carbonizes and turns to coke; and this is exactly what happens to the oil trapped in the turbo center bearing section after a hot shutdown. On the next startup, the turbo may not turn. An expensive trip to the shop is then called for.

Turbo coking can be avoided simply by running the engine at idle for four minutes or so after touchdown. (Actually, time spent rolling out on the runway and taxiing back to parking counts, so the wait is not as bad as one might think.) This procedure bathes the turbo shaft in relatively cool recirculating engine oil, and lets the turbo cool to non-coking temperatures before everything comes to a stop.

A brief cooldown idling period is not a bad idea for *any* engine, actually, and it may help prevent valve sticking in small Lycomings.

(See Lycoming Service Instruction No. 1425.) So before every shutdown, face your plane into the wind and idle the engine a minute or two with cowl flaps open (if any). You may confound and confuse impatient linepersons at strange airports with this technique, but in the end, which would you rather have: a happy lineperson, or a happy engine?

When the time comes to shut down, *first* turn off the boost pump, *then* pull the mixture control(s) into idle cutoff and let the prop come to a complete stop before turning ignitions switches off. The idea here, of course, is to keep excess fuel from accumulating in the cylinders (washing oil off your cylinder walls and perhaps gumming up your lower spark plugs) between flights, and also to reduce the chances of prop strike injury in the event your still-hot-engine's propeller is moved while pushing the airplane around on the ground. Killing the engine with the mixture rather than the spark also has the advantage of eliminating any chance of run-on due to hot-spot "dieseling." (If your engine "diesels" and fails to shut down quickly after moving the mixture into idle cutoff, take that as a warning symptom of internal carburetor or fuel-injector leakage. Or take it as a reminder to turn the boost pump off before shutting down!) You *could* kill the engine with the ignition, as in a car—and in fact you *should* do this every once in a while, to test for a broken P-lead (see Chapter 5)—but it's not recommended as a standard practice. If your car had a mixture control, killing *its* engine with ignition might not be standard practice, either.

Once the engine has come to a stop, be sure to turn the ignition off (pocket the keys) before doing anything else—again to reduce the likelihood of personal injury later if the prop is handled. Turn all electrical accessories off, then turn off the master switch itself. In Bendix pressure-carbureted aircraft, return the mixture control to about the halfway-in position; this prevents rubber diaphragms inside the carburetor from taking a set and affecting fuel flow on the next flight. Do *not* turn any airplane's fuel selector off between flights, unless fuel-system maintenance is to be done; it is too easy to forget to turn it back on at the start of the next flight.

One additional tip: If spark plug fouling has been a persistent problem, try running the engine to about 1,800 rpm (and holding it there 30 seconds or so) just prior to shutdown. (Return to idle momentarily for purposes of shutting down.) Spark plug cores scavenge well only at temperatures above 800 F, but at idle rpm those temperatures aren't reached, even with leaning. (Do lean during idle, though, to prevent carbon formation on plug electrodes.)

CHAPTER TEN
EMERGENCIES

Emergencies

No good statistics exist on the incidence of inflight powerplant emergencies in aviation. Accident statistics are available, but of course these tell us nothing about actual inflight shutdown rates, because not every engine stoppage results in an accident, obviously (and not every accident gets reported, in any case). Nevertheless, statistics kept by the National Transportation Safety Board show that—year after year—about one out of seven aviation accidents is caused by an engine malfunction not related to fuel starvation. In a typical year, U.S. civil aircraft rack up about 35,000,000 flight hours (or, accounting for twin-engined planes, about 40,000,000 *engine*-hours), incurring some 3,500 or so accidents, of which 500 are caused by engine problems. Thus it's possible to say that there is one NTSB-investigated engine malfunction every 80,000 engine-operating hours (approximately)—or, since the average engine goes 2,000 hours between overhauls, there's one engine-failure accident every 40 TBO runs. (Another way of looking at it is that since there are about 200,000 piston aircraft in the U.S., in any given 12-month period about one in every 400 piston airplanes will be involved in a powerplant-failure accident.) Many, if not most, engine stoppages are related to poor maintenance, so if equipment failures due to inadequate upkeep are removed from the above totals, the picture brightens considerably. (Indeed, accidents categorized by NTSB as due to "powerplant failure for undetermined reasons" account for only 250 mishaps per year, of which less than 10 percent are fatal.)

Of course, good or bad as these numbers are, there is room for improvement. A good place to start is with an understanding of how to manage an engine problem in flight. (Intelligent crisis management in turn demands a fair understanding of the basics of engine operation—which is why this chapter comes at the *end* of the book.) Bear in mind that the information given here is intended to supplement, not substitute for, advice given in your aircraft owner's manual and/or engine operating handbook. When a conflict exists, precedence should always be given to the information contained in your owner's manual or POH (Pilot's Operating Handbook).

General Principles

Remember that with very few exceptions, a light planes needs only 25 or 30 percent power to sustain flight, out of ground effect (somewhat less, very near the ground).* The goal, when managing any type of engine problem, is always the successful—preferably the on-airport—termination of the flight. Generally speaking, if through manipulation of the power controls you can get the engine to put out *some* power, chances are very good that you can make it to the nearest airport. (Even without power, chances are good for a survivable off-airport touchdown—*if* you continue to fly the airplane.)

Engine problems vary in severity, so it is hard to put forth any general advice on how to deal with an inflight malfunction. For purposes of this discussion, we will talk only of engine problems that (by their nature) call for immediate action: extreme roughness or vibration, severe power loss, gauge indications that exceed operating limitations, etc.—conditions that may not (strictly speaking) call for a "mayday" or formal declaration of emergency, but which are severe enough, nonetheless, to make the premature conclusion of the flight desirable.

No matter what the problem, always remember the following:

1. Fly the airplane. (Turn on the autopilot if you intend to devote full attention to solving the engine problem. However, if you are losing altitude, or if total engine stoppage is imminent, *do not engage the autopilot's altitude-hold feature.* Select attitude hold or simply trim the plane for minimum-descent speed.)

2. Reduce power. In almost any engine-malfunction emergency, the thermal and mechanical loadings on the engine can be reduced (and the likelihood of continued safe operation increased) by simply throttling back.

3. Do not continue flying with a sick engine. After a powerplant problem has been uncovered, plan on discontinuing the flight as soon as practicable. (A corollary is: Do not continue IFR with a sick engine if there is VFR weather not far away. Divert to the nearest VFR alternate, rather than make a low approach with an ailing powerplant.)

4. Do not continue the flight any longer than necessary after a serious problem has apparently "cured itself" or gone away. Intermittent roughness, or even a single loud (or unusual) noise, ought to be reason enough to discontinue the flight and begin an investigation into the cause.

As an aside, it has been the author's experience that engine prob-

---

* *Also recall that the typical small plane can glide two miles horizontally for every 1,000 feet of altitude AGL (above ground level).*

lems are more likely to be encountered immediately after maintenance than at any other time. Accordingly, it is poor practice to plan an IFR flight with passengers, under marginal conditions, immediately after any work is done on the plane's engine. (That includes annual inspections.) Consider the first flight after repairs to be a test flight.

Unexplained Engine Roughness

Engine roughness is arguably the most common, and hardest to troubleshoot, FWF (firewall-forward) complaint. One person's "loud noise" is another person's "resonant vibration." It is often difficult to tell prop or engine-mount problems apart from bonafide ignition or combustion aberrations. Nevertheless, certain guidelines apply.

First, observe the engine compartment for signs of damage (connecting rods sticking through the cowling, for example), oil or fuel leaks, smoke or flame, or anything at all unusual. If any unusual conditions are present, the most appropriate action will usually be to reduce power. (Of course, it goes without saying that if you're on fire, you should engage the extinguisher system if so equipped, or turn off the fuel at the fuel selector.)

Second, *consult the engine instruments.* If abnormal EGT, CHT, oil temperature, oil pressure, tach, or manifold pressure indications appear, proceed to the appropriate section below. (If fuel flow is fluctuating, engage the auxiliary fuel pump. Fuel-flow fluctuations are almost always caused by vapor bubbles in the lines.)

Third, attempt to reset the power controls in such a way as to make the engine run smoothly. If you're in icing conditions, select alternate air or carburetor heat, as appropriate. (Be ready for a large, temporary power loss if carb ice slushes off and enters the engine.) The throttle should be reduced, as mentioned above—with one consequence of this action being that a larger range of allowable rpms will be available (if the plane is equipped with a constant-speed prop). After throttling back and experimenting with prop rpm, *reset the mixture.* Keep all adjustments gradual, however. *Do not automatically shove the mixture control into the "full rich" position.* It's possible, especially in a turbocharged aircraft, or an aircraft with an especially powerful electric fuel pump, for the roughness to be the result of an already overrich mixture. Going to full-rich may well make the roughness worse.

After you've adjusted the power knobs (and assuming the roughness has not gone away), proceed to check out the ignition system. This means, of course, turning each magneto off one at a time to check for a defective harness or mag. Do this only after reducing power as described above. If the engine smooths out after one magneto has been shut off, *leave the bad mag off and continue to the nearest maintenance base.* There is nothing wrong with flying on one magneto in a situation such as this. (That's what you have two of them for.) The only problem is that in initially determining which—if any—of your mags is bad, you may well kill the engine when and if (in the process of troubleshooting) the faulty mag is selected. In the event you select one or the other mag, and the engine quits, *it is advisable that you momentarily reduce throttle and move the mixture to idle cutoff before turning the ignition back on.* (This will prevent a potentially damaging backfire.) Once the "good" mag has been found, of course, you can readjust throttle, rpm, and mixture as necessary to continue the flight.

It's possible, of course, that your engine may continue to run rough—or get rougher—after running through the checks outlined above. If the engine has sustained physical damage of a kind that prevents it from running smoothly, there may well be little or nothing you can do from the cockpit to prevent it from running poorly. This is true, for example, of valve-sticking episodes, where valves stick (and pushrods take a permanent bend) in flight. One also hears of valve lifters occasionally collapsing in mid-flight, producing a peculiar roughness that may or may not disappear later. This can be hard to troubleshoot.

Bear in mind, in any event, the following bit of advice, gleaned from the Continental TSIO-360-F *Operator's Manual* (p. 19): "WARN-ING: Severe roughness may be sufficient to cause propeller separation. Do not continue to operate a rough engine unless there is no other alternative."

High-Altitude Roughness

Roughness at (or going to) oxygen altitudes is usually a tipoff to ignition problems. There is a good reason for this. Magnetos do not generate a fixed voltage, but produce (by flyback effect) whatever output is, in effect, needed to jump the gap at the spark plug electrodes under any given set of conditions. Since air is electrically insulating, it takes more voltage to fire a spark plug under high-pressure conditions than low-pressure conditions. Conversely, it takes less voltage to spark a gap under a vacuum (or partial vacuum) than under standard sea-level conditions. (Ionization is encouraged in a vacuum, in other words.) The net effect, in a turbocharged engine, is that at some sufficiently high altitude—perhaps as low as 12,000 feet, perhaps

In a high-altitude airplane such as the Baron 58TC, engine roughness that goes away at low altitude is a tipoff to magneto or harness trouble.

20,000 feet or more (it depends on the plug gaps, plus many other factors)—a high-tension magneto will "spark out" internally rather than fire the spark plug. As one flies higher and higher, the atmospheric pressure in the magneto housing is less and less at the same time that there is a very high pressure in the combustion chamber. Electricity always takes the path of least resistance. Often that path is inside the magneto itself, at high altitude.

If misfiring is encountered at altitude, reduce manifold pressure. (Also try a different rpm.) Otherwise, *cruise at a lower altitude.* Continuing to fly with a misfiring engine is not acceptable in this case, even if the misfire is only occasional, as severe engine damage can occur.

Cleaning and regapping your spark plugs will help to forestall high-altitude ignition misfire (as will replacing any "leaky" ignition wires), but the ultimate answer to the problem is magneto pressurization. Magnetos pressurized with turbo bleed air are standard equipment on post-1981 Cessna T210, P210, and T303 aircraft as well as the Mooney 231, Piper's Navajo Chieftain and Malibu, Turboplus-modified Turbo Arrows and Senecas, and RAM-modified Cessna 340 and 414 aircraft.

Aside from magneto arcing, the other major cause of engine roughness at high altitude is fuel-flow fluctuation due to vapor formation in fuel lines (which can easily become heat-soaked after a long climb to altitude). If fuel-flow indications are fluctuating, energizing the boost pump(s) should rectify the problem. Some operating handbooks (such as those for the Cessna twins) advise leaving auxiliary fuel pumps on at altitudes above 12,000 feet, until engine temperatures stabilize in level cruise.

Major Power Interruption

Sudden, major power-loss events (with no prior warning from engine instruments) are rare, thankfully. Most such episodes stem from misfueling, or failure to remove water from sumps (occasionally vapor lock during a hot-day flight using substandard fuel); as a result, power is lost at fairly low altitude, with no time to go through checklists. We will confine our discussion here to enroute power interruptions, for which there is a sort of universal checklist (subject to some rather important caveats, if the engine is turbocharged; see below). For a *normally aspirated* engine, do the following at the first indication of major power loss:

1. Boost pump—ON (unless specified otherwise in owner's manual).

2. Fuel selector—SWITCH TANKS. (If you are already on the fullest tank, consider switching to another tank anyway. The first tank may not be venting properly, or it may contain condensation, bad fuel, etc.)

3. Mixture—RICH.

4. Alternate air—ON. (Turn OFF again if no improvement is noted.)

5. Magnetos—verify BOTH. (Try each mag independently if there is time.)

6. Primer—SECURED (in and locked).

7. Engine instruments—OBSERVE for indications of trouble. (If abnormal indications are found, refer to the appropriate sections further below.)

On an airplane equipped with a constant-speed prop, select high rpm. (There should be enough oil pressure for governing if the prop is windmilling.) The high rpm will, among other things, enable the magnetos to develop higher voltages to fire the spark plugs.

Always remember that if fuel, air, and spark are present, combustion *will* occur. The object of any "inflight power loss" checklist is to restore fuel, air, and spark to the engine, in quantities that will support combustion. If the checklist procedures fail to restore power, it may mean the engine has sustained physical damage of a type that precludes further operation. (For example, all oil may have been lost overboard, or the oil pump may have failed, causing bearing failure.)

As mentioned, this checklist is applicable to most lightplane engines, but is not to be used if it contradicts any advice given in the aircraft owner's manual. (Some owner's manuals advise turning the boost pump off immediately, for example, if turning it on briefly does not make the engine run better.) Also, as mentioned above, this checklist is for *normally aspirated engines only*. Turbocharged engines are subject to different procedures, outlined below.

Turbochager Run-Down

In a turbocharged engine operating at high altitude (above 12,000 feet), a transient interruption in fuel flow that might be acceptable in some other kinds of engines—such as might happen when (accidentally) running a fuel tank dry—can cause engine failure due to turbocharger run-down. (The power loss may or may not be complete, and may or may not be accompanied by surging of rpm, fuel flow, and/or manifold pressure.) In normal operation, a small turbocharger such as an AiResearch T04 or Rajay 325E10 reaches turbine speeds as high as 100,000 rpm. Any interruption of exhaust flow to the turbine will cause it to slow down immediately. At the same time, compressor output falls and a transient overrich mixture condition develops. If fuel flow to the engine is then increased or restored (perhaps by switching from an empty to a full tank, or activating the boost pump to eliminate vapor bubbles), the mixture becomes even richer—and before the turbine has time to catch up, the air/fuel ratio may fall below 8:1, at which point combustion ceases.

If your plane is turbocharged, follow the air-restart procedures in your owner's manual. These procedures vary significantly from plane to plane. If you do *not* have an air-restart checklist, do the following:

1. Mixture—IDLE CUTOFF.
2. Fuel selector—FULL TANK, or in a position that will permit the use of the electric boost pump.
3. Boost pump—ON (low position, if there is both a low and a high setting).
4. Throttle—OPEN to normal cruise position.
5. Propeller—HIGH RPM (cruise rpm or higher).
6. Mixture—ENRICH SLOWLY from idle cutoff. When engine restarts (indicated by a surge of power), continue to monitor fuel flow and make mixture adjustments as the turbocharger slowly comes up to full speed. Afterwards, readjust power levers and boost pump setting as required for cruise. Monitor manifold pressure closely for signs of overboosting.

If this procedure fails to result in a quick restart, it may be because you are at too high an altitude for normally aspirated combustion to want to occur. (Remember, above 12,000 feet the air is thin and you will get no more than a fraction of sea-level horsepower if the engine does restart. Also, you may be working against a closed wastegate, which creates significant exhaust back-pressure.) In other words, your chances of a successful "relight" increase at lower altitudes.

Turbocharger Failure

Catastrophic failure of turbochargers due to turbine imbalance, wastegate hangups, etc. are (thankfully) quite rare. When turbochargers fail (in small planes, at least), it's usually because of gas leakage through turbine housing cracks, rubbing of turbine wheel components against housing components, bearing failure, blade erosion, or damage from foreign objects (rivets, bolts, ball-point pens, etc.) being ingested into the compressor (or ingestion of valve heads, broken rings, piston aluminum, etc., into the exhaust turbine). Fortunately, these types of failures are rarely dramatic or life-threatening, with the exception of gas leakage through housing cracks, which can have a blow-torch-like effect on nearby accessories, cowlings, fuel lines, or oil lines. Barring an inflight fire (see discussion below), your first indication of a turbo-system failure will usually be an inability to get normal manifold pressure—lack of boost, in other words.

When a turbocharger failure occurs, the engine reverts, of course, to normally aspirated operation. But as explained in the section above, this does not mean the engine will run well—or put out anything approaching normal power. Most factory-turbocharged (non-retrofit-kit) engines incorporate low-compression-ratio pistons, for example, which have the effect of cutting sea-level horsepower substantially in the event of turbo failure. If your engine develops 285 horsepower (maximum) under normal circumstances, it may only develop 250 horsepower (at sea level) after turbocharger failure—or less, if the back-pressure of the (broken) turbocharger is excessive. Then too, at altitudes above about 12,000 feet, a turbocharger failure may well create an overrich mixture condition which can cause the engine to quit.

If a turbocharger failure is suspected due to a sudden loss of manifold pressure, readjust the mixture to obtain a fuel flow appropriate to the manifold pressure and rpm being indicated. Avoid any temptation you might have to respond to the power loss by shoving the mixture to "full rich." This is almost never a good idea in a power emergency in a turbocharged aircraft.

If manifold pressure falls to ambient pressure but the engine is not producing power (i.e., complete power loss at altitude), try the restart procedure give in your owner's manual, or (failing that) the one given further above, but first purge any excess fuel from the engine by retarding the mixture to idle cutoff and opening the throttle all the way. Then return the throttle and prop controls to their cruise positions, and proceed with a slow enrichment of the mixture. Land as soon as practicable.

Note: In any air-restart procedure, do not energize the starter or twist the key to the mag-retard (shower of sparks) position unless the propeller is stopped. If the magnetos are on, and the prop is turning the engine, attaining the proper air/fuel ratio in the cylinders (A/F between 8:1 and 18:1) will result in restarting.

At very high altitude, or on cold days, some difficulty in restarting a turbocharged engine may be experienced due solely to the frigid outside air temperatures. As Continental explains in its TSIO-360-F Operator's Manual, ''A few minutes' exposure to temperatures and airspeed at flight altitudes can have the same effect on an inoperative engine as hours of cold-soak in sub-Arctic conditions. If the engine must be restarted, consideration should be given to descending to warmer air (first).'' Allow a cold engine to warm up at 18 inches (approximately) of manifold pressure for several minutes before gradually increasing power.

Engine Fire

Inflight engine-compartment fires come about, typically, through breakage of old hoses and fittngs, failure of exhaust components, external fuel leakage from fuel injectors, arcing at chafed wires, and piston holing (which allows combustion gases to enter the crankcase directly, setting the oil on fire). Although no good statistics exist, fires are probably more prevalent in turbocharged-engine installations than unturbocharged systems; more prevalent in twins (with their tightly packed cowlings) than in singles; and perhaps more common in older planes than in newer ones. Good maintenance will go a long way toward preventing inflight fires (see *Light Plane Maintenance,* January 1981). So will an engine fire detection and extinguishing system.

Unless you know for sure where the fire is originating, it is a good idea to turn the master switch off (to eliminate any juice that may be causing arcing) as soon as possible after detecting smoke or flames. If you have a battery-powered portable comm radio, use it (and your ELT) to signal an emergency. Do not leave your panel avionics on any longer than necessary. (They will, of course, go off when you turn the master switch off.)

Pulling the mixture to idle cutoff will, obviously, cut off fuel at the flow divider or carburetor, but that may well be downstream of the fire's point of origin. To eliminate the fuel flow *upstream* of the combustion spot, *turn the fuel selector off.* (This is one of very few good times to turn a fuel selector off.) After that, you can pull the mixture back all the way—if in fact it will move at all (chances are, it's melted somewhere ahead of the firewall)—and turn the boost pump off (if it was on). Also secure the primer.

Very few owner's manuals give any advice at all on the use of cowl flaps in a fire emergency. Perhaps it doesn't matter. One's temptation might initially be to close them all the way to prevent blast air from fanning the flames. It depends on the location and type of fire, however. Usually, slowing the rush of air doesn't do any real good, because the resultant air/fuel ratio is still excellent for combustion. "Fanning the flames" makes a fire hotter only to the extent that it improves the air/fuel ratio for combustion; at a sufficiently high airspeed (i.e., with air in sufficient excess), the air/fuel mixture becomes too lean to support combustion and the fire goes out. (This is why a match goes out when you blow on it.) Accordingly, Cessna makes the recommendation in its Turbo 182 Owner's Manual that if, after going through the inflight fire checklist (which includes retrimming the aircraft for 100 knots) the fire has not gone out, the pilot ought to "increase glide speed to find an airspeed which will provide an incombustible mixture." (Another good reason to select a high glide speed: it gets you on the ground quicker.)

In summary, then, your inflight-engine-fire checklist ought to look something like this:

1. Master switch—OFF as soon as possible. (Call Mayday first, if you are going to.)

2. Fuel selector—OFF as soon as possible. (Do this before fooling with mixture or boost pump, etc.)

3. Mixture—IDLE CUTOFF (if it hasn't melted already).

4. Boost pump—OFF.

5. Primer—SECURED.

6. Cabin heat—OFF (to prevent smoke entry into cockpit).

7. Airspeed—ADJUST to minimize time in the air, and to find an airspeed that will provide an incombustible (or at least less combustible) mixture.

8. Cowl flaps—ADJUST to minimize flames/smoke.

9. Stick & rudder—SIDESLIP to keep flames away from cabin.

10. Gear & flaps—LOWER MANUALLY if fire has not gone out, prior to touchdown (at pilot's discretion).

11. Emergency exit—OPEN just before touchdown. (Many times, cabin entry doors and/or emergency exits get jammed shut on a hard landing as the fuselage deforms under load. Pop the latch before you hit, so you can get out.)

If the fire self-extinguishes before touchdown, you might be tempted to turn the master switch back on (to lower the gear, for example). It's probably a bad idea, especially if one or more circuit breakers has popped.

Low Oil Pressure

Pressure oil lubrication is what keeps a reciprocating engine reciprocating. (It is also, in part, what keeps an air-cooled engine cool.) Total loss of oil pressure, whether because of oil-pump failure, oil starvation, or for other reasons, can have dramatic consequences. At some point, bearings seize and—if the pilot is lucky—the prop simply stops cold. If the pilot is not so lucky, a connecting rod comes through the windshield and causes dental problems.

This is not to say, of course, that an engine won't continue operating with oil pressure just below the green arc. Lycoming and Continental vary in the way they locate their oil-pressure pickoff points (see Chapter 2), so what's true for one type of engine isn't necessarily true for another; but pilots tend to underestimate an aircraft engine's ability to run with low oil pressure *at low power settings*. High crankshaft speeds, high temperatures, and high brake mean effective pressures put a high demand on lubricating oils. Reduce the demands, and the oil's "operating envelope" is expanded somewhat. As long as temperatures and power settings (and rpm) are not high, many engines will drone along without damage at oil pressures as low as 10 or 20 psi. Most engines will idle for at least short periods, without damage to bearings or lifters, at pressures less than 10 psi. (A lot depends on where and how you measure the pressure, however, and the grade/quality of oil.) A full-power go-around requires significant oil pressure, on the other hand (as do turbo bearings and wastegate actuators).

Bearing all of this in mind, the very first thing to do when low or fluctuating oil pressure is noted (fluctuation usually signalling oil-pump cavitation or imminent oil exhaustion) is reduce power.

The very second thing to do after noting a low oil-pressure indication is look at the oil-temperature gauge. When oil flow through the engine galleries slows down (as it inevitably must when pressure falls off), oil temperature invariably rises. (The rise in temperature thins the oil and reduces pressure even further.) If high oil temperature is noted, open the cowl flaps and trim for a higher airspeed. Since you've already reduced power, this means going down. For high-flying aircraft, this is definitely a good idea, since engine cooling is better at lower altitudes due to the denser air. (It's true that air is colder at high altitudes, but the decrease in temperature is more than offset, for engine-cooling purposes, by the decrease in air density.) Since oil temperature is high, you can be sure that you have a bonafide oil-pressure problem requiring immediate attention. Land as soon as possible.

If oil temperature is *not* high and you have a low pressure indication, suspect a bad gauge rather than a bonafide oil-pressure problem. (If oil is dribbling onto the cabin floor beneath the pressure gauge, you can be certain the problem is with the gauge!) The oil pressure gauge plumbing contains a restrictor orifice designed to prevent rapid loss of oil from the sump in case of a plumbing failure; hence, you are in no immediate danger if the problem is leaky plumbing. Then too, it often happens that a tiny bit of carbon or sludge will block the orifice in the gauge plumbing, causing the cockpit instrument to indicate erratically. Problems of this sort are by no means uncommon. The tipoff is always low pressure in conjunction with normal oil temperature indications (and normal engine behavior). Continued flight is permissible, provided gauges are monitored closely for changes.

When oil pressure falls into the "limbo zone" just below the bottom of the green (but above, say, 25 psi), with oil temperatures high but not at redline, it is usually advisable to continue the flight to the nearest convenient maintenance base. Pressure drops of this magnitude, particularly if occurring shortly after a change to a new brand of oil or following an engine overhaul, are often found to be attributable to bits of carbon (washed out by the oil) hanging up on the oil pressure relief valve inside the engine. This spring-loaded valve acts in popoff fashion to prevent oil pressure excursions on the high side. If dirt, metal, carbon, washers, gasket materials, etc. become lodged under the valve, preventing it from seating, pressure oil from the oil pump is simply dumped back to the sump, and low oil pressure is seen in the cockpit. Frequently, the carbon works free by itself, and the problem goes away without a mechanic's help. (Often it doesn't. So have an A&P look at the valve before flying the engine again.)

Any time oil pressure falls into a red or yellow arc on the pressure gauge (or below 20 psi, in general)—accompanied by a sharp rise in oil temperature—there is some danger of catastrophic engine failure, and the pilot should plan accordingly. If pressure is extremely low (and oil temperature high), a precautionary landing should be made—off-airport, if necessary. Do not ignore the warning signs.

Needless to say, if pressure is low and oil is streaming from the engine compartment, total engine failure can be considered imminent. (Don't count on Teflon additives to save the day.) Pick a spot and land.

High Cylinder Head Temperature

When the CHT gauge reads high, it's sometimes difficult to know whether you have an overtemp condition merely in one cylinder (i.e., the one to which the sensing probe is attached) or all cylinders. EGT does not give a reliable crosscheck, for reasons described in Chapter 8. (CHT can increase as EGT goes down, and vice versa.) Nor can you assume, when the CHT gauge is showing normal cylinder temperatures, that the temperatures of your other cylinders (the ones to which the probe is *not* attached) are not outside limits. CHT indications must therefore be treated with a bit of caution.

Nevertheless, a CHT indication in the high 400s (Fahrenheit) is usually a reliable tipoff to trouble. CHT redlines are variously set at 460 and 500 degrees. Anything above 425 can be considered

Cylinder cooling is critical in a high-output engine. But it pays to remember that in most planes, there is only one CHT probe. CHTs above 425 F should be avoided.

179

detrimental to the long-term health of the cylinder head, and anything approaching redline is cause for genuine concern. The answer is to reduce power, enrich mixture (to the extent that it is possible to do so without causing engine roughness), open any cowl flaps, and—if the aircraft is at high altitude (above 12,000 feet)—try a lower cruise altitude. If it is practical to do so, the airplane should also be trimmed nose-down for a higher airspeed.

Of course, reducing power in level flight reduces airspeed, too, which would seem to be a poor thing to do in the event of CHT overtemping. In fact, though, an airplane's engine is much better at producing heat than propulsive power (airspeed). It takes a very large increase in power to get a relatively small increase in airspeed; and the reverse is also true. By throttling back, you cut heat production dramatically, at the expense of very little loss in cooling airflow.

Going to a lower altitude may sound counterintuitive as a remedy for high CHT (after all, the air is much cooler at altitude, isn't it?), but the fact is that air *density*—which is very important for cooling—falls off so rapidly at higher altitudes (along with indicated airspeed and cooling drag) that the low outside air temperature doesn't compensate. Cooling is worse up high than down low.

Usually an abnormally high CHT indication can be brought down to acceptable levels by a combination of mixture enrichment, opening cowl flaps, and/or reducing power or altitude. When it can't, you're more than likely looking at one or more of the following:

—a partially clogged injector nozzle (which will cause a lean-out in just one cylinder);

—deteriorated engine cooling baffles;

—an induction air leak;

—advanced mag timing

—fuel pump or injector system or carburetor not set up within specified fuel-flow limits;

—preignition;

—and/or detonation.

Bad fuel will cause detonation, as will aggressive leaning at high power settings (or even at moderate power settings, at high altitude, in some turbocharged engines), or for that matter sufficiently advanced timing. If you have just had your magnetos worked on, suspect the latter. Regardless of the cause, if detonation is suspected, reduce power and enrichen the mixture.

In some instances, high CHT may accompany valve-burning. If a valve stops rotating (due to deposit buildup in the neck area, for example), hot gas leakage can produce a high head temp.

High Oil Temperature

High oil temperature and low oil pressure go hand in hand. If you have one, you should eventually also have the other. Therefore crosscheck the two gauges, and before totally "believing" either one, see if it agrees with the other. Prolonged high oil temperature, not accompanied by a significant (several needle widths) dip in oil pressure, may be indicative of nothing more than a faulty capillary or thermocouple, which is quite common (see Chapter 2). The old-style capillary-type sensing systems were and still are quite susceptible to handling damage in the course of maintenance.

The conditions under which high oil temperature manifests itself can be important in diagnosing the source of the problem. For example, many engines have a so-called vernatherm valve at the oil cooler which acts thermostatically to shunt oil away from the cooler any time the temperature demands on the oil are not large. Hangups in this valve are quite common, and give rise to a predictable syndrome. Oil temperature typically builds and builds after takeoff, reaching near-redline levels (where it may stay for quite some time) before—suddenly and for no apparent reason—plunging to the low green. This is the tipoff to a sticky vernatherm. Any time wildly fluctuating oil temperature is noted, the thermostatic bypass valve (or vernatherm) should be taken out and inspected. In some cases, it is advisable to put a thicker or thinner gasket under the vernatherm valve, to increase or decrease the distance the valve has to move (thereby altering its sensitivity).

If chronic (not fluctuating) high oil temperature is a problem, and engine compression is normal during a differential-compression test (ruling out blowby as the cause), calibrate the temperature-sensing probe with boiling water, and if the indicator system is working properly, check the oil cooler and associated plumbing. (Oil coolers occasionally become clogged, cooling fins become bent, etc. Depending on the design and location, coolers also can become vapor-locked with air bubbles trapped in the high spots in the system.) It's also a good idea to check engine cooling baffles and general cowling conditions, to make sure the oil cooler is getting the proper amount of blast air. If, after checking all the "easy" things, you still haven't pinpointed the cause of chronic high oil temperature, pull the accessory case cover and check the oil pump.

The grade of oil used—and the amount—will have an effect on oil temperatures, naturally, as will ambient conditions. While most aircraft engines can be successfully operated (from a pure lubrication standpoint) in normal flight attitudes with as little as two quarts of oil on board, from a *cooling* standpoint it is desirable never to dip below the minimum level specified in the airplane owner's manual (usually six quarts).

In any event, the proper course of action when high oil temperature is noted in flight is, of course, to throttle back (and open the cowl flaps). If the redline temperature has not been exceeded, there is no immediate danger in continuing the flight as long as oil pressure remains normal.

Overspeeding

Crankshaft overspeeding is normally not possible in a fixed-pitch prop airplane unless the aircraft is in an unusual attitude (or doing aerobatics), or redline airspeed is being approached. Overspeeding is more commonly observed in constant-speed-propeller airplanes, where the condition may be caused by governor malfunction, pilot-induced surging, or overboosting (exceeding manifold-pressure limits in a turbocharged airplane). Whatever the cause—and whatever the type of plane—corrective action consists of immediately reducing power with the throttle and (if it is working) the prop pitch control. It may also be necessary to raise the nose. (Here again we see an instance in which the elevator can function as a powerplant control.) In a steep nose-down attitude, even reducing power to idle may not prevent an overspeed condition. It depends on the airspeed.

Lycoming defines "momentary overspeed" as operation for not more than three seconds at rpms not exceeding 110 percent of rated rpm. (Use the takeoff rating, if the engine has both a max-continuous and a five-minute takeoff rating.) Any operation at more than 110 percent, for any duration, should mean an immediate oil change, oil spectrum analysis to detect severe wear or imminent component failure, borescope inspection of cylinders, visual inspection of pushrod ball ends and tappet faces, visual check of valve tips and rockers, and a compression check (plus examination of oil filter or screen contents for metal). Overspeeding beyond 120 percent of rated rpm, for any duration, calls for special action (i.e., engine disassembly and wholesale replacement of reciprocating parts); see Lycoming Service Bulletin No. 369 and/or Continental Service Bulletin M75-16 for details.

GLOSSARY

Glossary

A/F ratio
air-to-fuel ratio, by mass.

AGL
Above ground level.

angle-valve head
Any cylinder head in which the intake and exhaust valves are not parallel. Angle-valve heads ''breathe'' more efficiently than parallel-valve heads. Most (but not all) currently produced aircraft engines are of the angle-valve-head type.

austenitic
Refers to any iron alloy containing carbon.

automatic controller
A device, usually employing an aneroid bellows, used to control the opening and closing of a turbo wastegate (see *wastegate* below). Automatic controllers come in various configurations, but the most common type is the absolute pressure controller. The controller is similar to a large barometric altimeter, referenced to deck pressure (see *deck pressure* below) instead of the atmosphere. When the aneroid bellows expands or contracts, it repositions a needle valve or oil poppet controlling oil pressure to the (hydraulic) wastegate actuator. Thus, a constant, absolute pressure is maintained in the induction system ahead of the throttle. Usually, automatic controllers have no mechanical connection whatsoever to cockpit power levers, but respond *indirectly* to manifold pressure demands created by throttle movement.

babbitt
An alloy of tin, copper, and antimony often used to coat bearing surfaces.

bhp
Brake horsepower—the actual amount of horsepower delivered from an engine's crankshaft to a propeller or test-stand brake.

bifilar pendulum damper
A type of crankshaft counterweight which hangs on two roller pins and is free to swing back and forth on the parallel-rolling pins. Crankshaft dynamic counterweights are of the bifilar configuration.

bmep
Brake mean effective pressure. A measure of the average pressure (in psi) developed within a cylinder, based on actual brake horsepower, rpm, and displacement. Bmep is directly proportional to power output, and inversely proportional to displacement and rpm. (E.g., if two engines of identical displacement are producing the same power output, but at different rpms, the slower-turning engine is operating at a higher bmep.)

boost
Extra manifold pressure resulting from turbocharger output.

boost pump
Auxiliary electric fuel pump.

bootstrapping
A condition which can occur in turbocharged aircraft in high-altitude cruise (with the wastegate closed—see *wastegate* below), in which large, unexpected manifold-pressure excursions take place spontaneously. The manifold-pressure instability is due to feedback-loop effects in the turbo system. The usual "cures" are to increase engine rpm and/or open the wastegate.

borescope
Device for making a detailed *in situ* optical inspection of remote engine parts.

Bourdon tube
A miniature inflatable, coiled tube which is used in many types of instruments to drive a meter movement.

Bowden cable
A type of cable in a sheath, commonly used for throttle runs, prop controls, mixture, etc.

breather
Crankcase breather tube—a metal tube that runs from a point on the upper surface of the crankcase (often near the front of the engine) to an exit hole in the bottom of the cowling. The purpose of the

breather is to vent crankcase vapors which might otherwise collect and cause crankcase pressurization.

BTC

Before top center (of piston travel).

BTU

British Thermal Unit. A common unit of energy; specifically, the amount of energy needed to raise the temperature of a pound of water one degree Fahrenheit.

butterfly valve

The pivoting baffle plate (or air valve) found in the throttle body area of a conventional fuel injector or carburetor. The throttle is connected (by a long cable) directly to this valve. Opening and closing of the butterfly directly controls the amount of air admitted to the engine, whether it is a turbocharged or normally aspirated engine.

CAFE

Competition for Aircraft Flying Efficiency. An aircraft efficiency race that was held yearly in California starting in the 1970s. The object of the race (open to all forms of powered, heavier-than-air aircraft) is to achieve the highest score based on a formula that multiplies payload, speed, and miles-per-gallon fuel efficiency together.

choke

In an aircraft engine, choke refers to the slight bit of taper machined into the cylinder barrel at the top of ring travel. The taper is necessary to ensure that the barrel has a perfectly cylindrical contour at normal operating temperatures.

CHT

Cylinder head temperature. The temperature of the large, finned end of the cylinder containing the valves.

compression ratio

The ratio of the combustion-chamber volume (the space above the piston) before and after the piston reaches the extreme limits of its travel. The Continental O-470-R is a typical "low compression" engine, with a C.R. of 7.0-to-1. (By comparison, many radial engines were 6.0-to-1 or 6.5-to-1, while the highest compression ratios are seen in the Lycoming O-360-E and O-320-H families, where a C.R. of 9.0-to-1 is used.) High compression ratios permit efficient power production but generally require the use of higher-octane gasolines.

creep
The tendency of a metal to become plastic and flow at high temperatures.

critical altitude
In a turbocharged aircraft, the altitude above which sea-level rated manifold pressure can no longer be maintained; or the altitude above which the *wastegate* is closed all the way, all the time.

deck pressure
A term used in conjunction with turbocharged engines. Refers to the pressure in the induction system between the turbo compressor outlet and the throttle butterfly. This is not the same as manifold pressure. Manifold pressure (see entry below) is always measured downstream of the throttle butterfly. In a turbocharged engine, some excess of pressure in the deck area—ahead of the throttle plate—is desirable so that when the pilot advances the throttle, the engine doesn't hesitate.

detonation
Combustion knock. It can be defined as the premature, spontaneous auto-ignition of the unburned fuel-air charge ahead of the flame front, in a combustion chamber in which sparking of the spark plug(s) has already occurred. In other words, the spark event has taken place at its normal time, and fuel and air are present, and a portion of the fuel and air has begun burning; but the unburned portion (compressed by the expansion of the burning charge) reaches a pressure and temperature sufficient to cause a sudden, all-at-once explosion of the entire charge. This produces the familiar knocking sound in an automobile engine. It also produces mechanical stresses which can eventually fail rings, pistons, connecting rods, or cylinder heads or valves. In an airplane engine, the pinging or knocking sound cannot usually be heard from the cockpit. Hence it is especially important for a pilot to avoid letting an engine detonate in the first place. Detonation can be caused by fuel of insufficient octane rating, or (with the proper octane fuel) operation a too high a power output with too lean a mixture. Improperly advanced magneto timing will also hasten the onset of detonation, as will heating of the incoming fuel-air charge (by carburetor heat or by compression via turbocharger). Compare ''preignition.''

displacement

The cylinder cross-section area multiplied by the piston travel (stroke), multiplied by the number of pistons in the engine. Displacement, in cubic inches, is a handy means of comparing engine sizes, although one must be careful to remember that horsepower does not correlate directly with displacement, since such factors as compression ratio (see entry above) and volumetric efficiency (see entry below) also affect power output.

dynamic counterweights

Pendulum-type weights which are used in opposed pairs on the crankshaft to ensure dynamic balance and dampen out potentially damaging resonant vibrations at critical rpms. Not all aircraft engines have dynamic counterweights (most four-cylinder engines do not, for example).

EGT

Exhaust gas temperature, or an instrument that indicates same.

F/A ratio

Fuel-to-air ratio, by mass.

float carburetor

A low-pressure carburetor employing an integral reservoir or bowl (from which fuel is siphoned into the engine), the level of fuel in which is controlled by a valve on an arm connected to a float, much like the action inside a water closet. Float carburetors for aviation are manufactured by Facet Aerospace (Jackson, Tennessee) under the Marvel-Schebler brand name. An alternative to the float carburetor is the *pressure carburetor*, which has no bowl but meters fuel under moderately high pressure (from a fuel pump) to a jet or jets near the throttle butterfly, not unlike automotive throttle-body injection. (Pressure carburetors were made by Bendix-Stromberg in the 1950s and are found on many Eisenhower-era planes.)

flow divider

In a fuel-injection system, the fitting (atop the engine) into which the individual cylinder fuel-delivery lines connect; the point at which fuel flow is divided so as to allow arrival of fuel at the individual cylinders. Also called the "injection spider" or (in Continental parlance) the "manifold valve."

four-stroke cycle

Also called *Otto cycle*. This is the familiar power production cycle in which fuel and air are drawn into the cylinder on the piston downstroke; the fuel/air charge is compressed on the next piston upstroke; combustion takes place on the subsequent downstroke; and exhaust gases are expelled from the combustion chamber on the following upstroke. All current production aircraft engines are of the reciprocating (vs. rotary/Wankel), spark-ignition (vs. diesel), four-stroke-cycle (vs. two-stroke cycle) type.

FWF

Firewall-forward.

gallery

Oil passageway in crankcase.

Heli-coil

Trademark name for a brand of helically coiled thread inserts. The coil metal has a diamond cross-section so that, once installed, the Heli-coil itself becomes the ''threads'' for the hole or boss in question. Heli-coils are used not only for repair of oversize threads but as original linings for spark plug (and other) holes/bosses.

IAS

Indicated airspeed, as measured by a pitot-static type airspeed indicator system.

Inconel

Trade name for a family of high-nickel superalloys resistant to corrosion at high temperatures. A few exhaust valves and some exhaust pipes/clamps are made of Inconel. (Because of its good material properties at high temperatures, Inconel was used extensively in the manufacture of the X-15 rocket plane.)

induction system

Broadly, the entire air-intake system of the engine, from air filter (or airscoop) to intake ports.

intake manifold

The system of piping by which air is delivered to the cylinders.

intercooler

An air-air radiator placed in the induction system of a turbocharged engine, between the turbo compressor outlet and the throttle butterfly. The purpose of the intercooler is to cool down the intake air (which may emerge very hot from the compressor outlet).

manifold pressure

The pressure, in inches of mercury, of air within the engine induction system downstream of the throttle butterfly. This pressure is directly related to air flow and hence (under most circumstances) engine power output.

manifold valve

—see *flow divider*.

MP

Manifold pressure.

MSL

Mean sea level.

Nimonic

Trade name for a family of nickel alloys designed for high-temperature applications. Nimonic 80A is used in virtually all Continental exhaust valves, and many Lycoming valves as well.

Ni-Resist

A heat- and corrosion-resistant nickel-iron alloy, often used, for example, in valve guides and turbocharger housings.

Nitralloy

A nitride-hardened steel alloy.

normally aspirated

Any engine that does not employ supercharging or turbocharging is said to be *normally aspirated* because it relies purely on piston action to pump (draw in; aspirate) air for combustion.

normalize

Restore manifold pressure to sea-level equivalency via turbocharging. Most aftermarket (STC'd) or ''add-on'' (bolt-on) turbocharger installations are intended to provide normalization of manifold pressure; that is, redline MP remains at 29 inches (or whatever it was before turbocharging), and the turbo wastegate is operated in a such a way as to restore sea-level manifold pressure in the climb. By contrast, most factory turbo installations are intended to *ground-boost* the engine to manifold pressures significantly higher than 29 inches.

NTSB
National Transportation Safety Board.

OAT
Outside air temperature. The temperature of the atmosphere outside the airplane.

Otto cycle
—see *four-stroke-cycle* above.

overboost
Broadly, any condition which has an engine operating in excess of max rated power. In a turbocharged engine, it refers to operation beyond redline manifold pressure.

P-lead
Magneto primary lead wire.

PMA
Parts Manufacturer Approval. A type of approval granted to manufacturers by FAA.

preignition
Ignition of the fuel-air charge in a cylinder *before* the normal discharge of the spark plug. In a broad sense, anything that causes combustion to occur before the specified ignition timing (as set forth on the engine data plate) could be said to be causing preignition; thus, even the aircraft ignition system—if maladjusted—could cause preignition to occur. Because of the abnormal stresses induced, preignition can be extremely damaging to an engine, and at high power settings the damage can occur very quickly (often in less than 30 seconds). Operation with detergent-type automotive oils can cause preignition due to ash deposits (from the barium- and calcium-containing detergents) which leave hot spots in the combustion chamber. (Hence the use of "ashless dispersant" oils in aviation.) If piston or rod failure doesn't occur first, preignition will make itself evident by very high CHT indications.

pressure ratio
In turbocharging, the ratio of outlet to inlet pressure at the turbo compressor. Example: At 18,000 feet, where the ambient atmospheric pressure is 15 inches of mercury, a turbocharged engine operating at 30 inches of manifold pressure is operating with a turbo-compressor *pressure ratio* of 2.0-to-1 (minimum).

psi

Pounds per square inch. A measure of pressure. (One psi is approximately equal to two inches of mercury.)

Rankine

A temperature scale starting at absolute zero, but using Fahrenheit-type degree divisions. (The equivalent scale using Celsius-type degrees is called the Kelvin scale.) Since absolute zero occurs at minus-460 Fahrenheit, the freezing temperature of water is 492 Rankine. Many engineering calculations involving compression or expansion of gases require conversion of all temperatures to Rankine numbers.

Reid vapor pressure

A measure of the tendency of a fuel to form vapor bubbles.
Industry specifications for automotive gasoline and avgas differ somewhat with regard to allowable Reid vapor pressure. At present, the maximum for avgas is 7.0 psi, while the maximum RVP for winter auto fuel is on the order of 12 psi.

SAE

Society of Automotive Engineers. Oil viscosity is measured via a method worked out by SAE, hence 'SAE 30,' 'SAE 40,' etc. to describe grades of oil.

sfc

Specific fuel consumption, usually given in pounds of fuel per hour per horsepower (lbs/bhp/hr). Sfc is a measure of the fuel efficiency of an engine with regard to horsepower production. For aircooled gasoline engines, sfcs of 0.5 to 0.6 are considered typical.

STC

Supplemental Type Certificate. This is a type of authorization required by FAA before certain accessories (such as turbocharger kits, special propellers, etc.) can be added onto an airplane or substituted for original equipment.

stoichiometric

Refers to conditions in which every reactant in a chemical reaction is used up completely, with no reactants left over. Thus, a stoichiometric fuel-air mixture is one in which the ratio of oxygen to gasoline is perfectly balanced so that neither is present in excess. Combustion of a stoichiometric mixture leaves nothing but carbon dioxide and water (and heat).

supercharger
Broadly speaking, any device that forcibly increases the amount of air flowed by an engine's induction system. Although technically speaking a *turbocharger* is a type of supercharger, today the word supercharger is usually used to connote a *mechanical supercharger* (that is, a fan or pump driven mechanically off an engine's crankshaft). Mechanical superchargers can be found on many radial engines, and on certain 480- and 540-cubic-inch Lycoming engines used (for example) on Beech Queen Airs and some Aero Commanders.

TBO
Time between overhauls.

throttle butterfly
—see *butterfly valve*.

TIT
Turbine inlet temperature. The temperature of the exhaust going into the turbine portion of a turbocharger.

turbocharger
Also sometimes called a "turbosupercharger." A turbocharger is an exhaust-driven supercharger (see "supercharger" above), used for increasing the manifold pressure—and therefore the total power output—of an engine. Aircraft turbochargers are made by Roto-Master (under the brand name "Rajay") and Garrett's AiResearch Industrial division.

upper deck
The portion of a turbocharged engine's induction system upstream of the throttle butterfly and downstream of the turbo compressor outlet. (See also *deck pressure*.)

venturi
An hourglass-shaped restriction in a tube, pipe, or other conduit. When a fluid is pumped through a venturi, flow speed increases at the point of maximum taper. Pressure also decreases at this point (by Bernoulli's law). In carburetors, as well as some types of fuel injection systems (such as Bendix RSA injectors), a venturi is used in the throttle body section, with impact tubes at various points to measure the pressure drop in the venturi (which is proportional to

flow rate). Fuel is then metered in proportion to the pressure drop.

vernier

A type of push-pull control knob which can either be pushed and pulled directly, or adjusted very finely by twisting. Vernier controls are most commonly used to control prop pitch (or prop rpm). They are also used for manual wastegate turbo systems, and—less frequently—for throttle or mixture control in certain airplanes.

volumetric efficiency

The ratio of the *actual* amount of air passed through a cylinder (or an engine) in one power cycle—i.e., two crankshaft revolutions, in a direct-drive, Otto-cycle engine—to the cylinder's (or engine's) displacement. Volumetric efficiency is thus a measure of how well an engine "breathes." For a normally aspirated aircraft engine, a V.E. of 70 or 80 percent is considered normal. Volumetric efficiencies of over 100 percent are possible with turbocharging or supercharging.

Engine Nomenclature

Lycoming and Continental have for the most part wisely avoided naming their engines after racehorses and icons (the exception being Continental's ill-fated Tiara series), instead adhering to a simple—and informative—shorthand based on configuration and displacement. For example, O-200 means "opposed cylinders, 200 cubic inches displacement." Similarly, IO-520 means an engine that is injected, opposed, and has 520 cubic inches' total displacement. (When there is no letter 'I' in the designator, meaning no injection, the engine is assumed to be carbureted.) As it happens, there are many different models in, for example, the IO-520 family, differing significantly in horsepower, rpm, accessory layout, and even cylinder and crankcase design. These differences give rise to the first letter suffix (IO-520-A, IO-520-B, -C, etc.) in a given engines series. Unfortunately, there is no easy way to tell from the simple notation "IO-520-B" what the engine's horsepower rating is, or how it differs from an IO-520-C, IO-520-D, etc. (This information can be obtained, however, by referring to the FAA's Type Certificate Data Sheets, which you can buy from the Government Printing Office or perhaps borrow from a willing A&P.)

Turbocharged engines carry an extra prefix in the name designation: 'TS' (for turbosupercharged) if Continental, 'T' (for turbocharged) if Lycoming. Thus Continental's basic turbocharged engine families are the TSIO-360 and TSIO-520; Lycoming has the TO-360, TIO-540, and several others. Why Continental prefers the old-fashioned term "turbosupercharged" to the more succinct (and synonymous) "turbocharged" is anyone's guess.

Engines with prop reduction gearing carry a 'G' in the prefix, in addition to everything else. Hence GTSIO-520 (Continental) and TIGO-541 (Lycoming). Some of these engines, such as Continental's GTSIO-520 series, carry a single large spur-type reduction gear at the front of the crankcase; others, such as the (now out of production) GO-435, GO-480, and IGSO-540 Lycomings, have planetary reduction gearing. There is no way to tell from engine nomenclature, however, which type of gearing is used on a particular engine; you simply have to remember which engine is which.

Some engines, such as the GSO-480 and IGSO-540 Lycomings (both out of production) are *mechanically supercharged*—signified by an 'S' in the prefix.

The commonly used prefixes can be summarized as follows:

Prefix

A	Aerobatic (inverted oil system, etc.).
G	Gear reduction prop drive.
H	Helicopter model, horizontal orientation.
I	Fuel-injected.
L	Lefthand engine rotation.
O	Opposed cylinders.
S	Supercharged (Lycoming).
T	Turbocharged (Lycoming).
TS	Turbosupercharged (Continental).
V	Vertical orientation (helicopter).

The first letter in a suffix (the 'C' in IO-520-CB, for example) generally doesn't stand for anything in particular but merely serves to differentiate between models in a given displacement series. When the suffix is more than one letter long, the second letter (or number) frequently does have a standard meaning. In Continental engines, for instance, a 'B' following the first suffix letter (as in TSIO-360-EB, TSIO-360-FB, etc.) signifies the presence of an improved, B-series crankshaft. (All 360- and 520-series Continentals made after 1979 incorporate the improved crank.) Confusing matters slightly is the fact that a few years ago, Continental ran out of letters in the alphabet to name its TSIO-520 engines and so resorted to a dual-letter basic suffix ending in 'E': for example, TSIO-520-AE, TSIO-520-BE, TSIO-520-CE.

Lycoming suffixes are often up to five designators long—for example, O-540-L3C5D. Here, the 'L' in -L3C5D merely designates the basic engine subfamily, as before. The other letters and numbers are primarily of interest to factory technicians. The '3' denotes the type of nose section used; the 'C' refers to the accessory section configuration; the '5' refers to dynamic counterweight configuration; and the 'D' means the engine is equipped with a Bendix dual magneto (i.e., a 2000-or 3000- series Bendix mag containing two rotors and two distributors, and two sets of points, driven off a single shaft). Whenever you see a 'D' at the very end of a Lycoming engine designation, it means a single-drive dual mag is in place. (Since no one but Bendix makes a single-drive dual mag, you can automatically assume the mag is a 2000-or 3000-series Bendix.) Thus, an IO-360-A1B6D has one Bendix 2000-series mag; an IO-360-A1B6 has two independently-driven magnetos (which may or may not be Bendix).

Continental has chosen to forgo the laborious Lycoming system of letters and numbers in the final designator, instead preferring to use a "spec number" to alert factory (and overhaul) technicians to the accessory configuration of the engine. On a Continental engine data plate, then, you might see the complete engine name given as "TSIO-360-FB, Spec. 16." The 16 in this case is useful to factory people in determining the exact type of fuel injector used, the type of accessories in place (for example, the model of magneto and whether or not it is pressurized), and often a variety of other fine points. The number is of no particular significance to pilots.

Standard Atmosphere Chart

Altitude (ft.)	Air Pressure (in.Hg)	OAT (F)
S.L.	29.92	59.0
1,000	28.86	55.4
2,000	27.82	51.9
3,000	26.81	48.3
4,000	25.84	44.7
5,000	24.90	41.1
6,000	23.98	37.6
7,000	23.09	34.1
8,000	22.23	30.5
9,000	21.39	26.9
10,000	20.58	23.4
11,000	19.80	19.8
12,000	19.03	16.2
13,000	18.3	12.7
14,000	17.58	9.1
15,000	16.89	5.6
16,000	16.22	2.0
17,000	15.58	-1.6
FL 180	14.95	-5.1
FL 190	14.35	-8.7
FL 200	13.76	-12.3
FL 210	13.20	-15.8
FL 220	12.65	-19.4
FL 230	12.12	-22.9
FL 240	11.61	-26.5
FL 250	11.12	-30.1
FL 260	10.64	-33.6
FL 270	10.18	-37.2
FL 280	9.74	-40.7
FL 290	9.31	-44.3
FL 300	8.90	-47.8

INDEX

Index